Six-Minute Solutions

for Civil PE Exam Structural Problems

Second Edition

Christine A. Subasic, PE

Professional Publications, Inc. • Belmont, CA

How to Locate and Report Errata for This Book

At Professional Publications, we do our best to bring you error-free books. But when errors do occur, we want to make sure you can view corrections and report any potential errors you find, so the errors cause as little confusion as possible.

A current list of known errata and other updates for this book is available on the PPI website at **www.ppi2pass.com/errata**. We update the errata page as often as necessary, so check in regularly. You will also find instructions for submitting suspected errata. We are grateful to every reader who takes the time to help us improve the quality of our books by pointing out an error.

SIX-MINUTE SOLUTIONS FOR CIVIL PE EXAM STRUCTURAL PROBLEMS
Second Edition

Current printing of this edition: 1

Printing History

edition number	printing number	update
1	1	New book.
1	2	Minor corrections.
2	1	Code updates and minor corrections. Copyright update.

Printed in the United States of America

Professional Publications, Inc.
1250 Fifth Avenue, Belmont, CA 94002
(650) 593-9119
www.ppi2pass.com

Library of Congress Cataloging-in-Publication Data
Subasic, Christine A., 1966–
     Six-minute solutions for Civil PE exam structural problems / Christine A. Subasic.
          p. cm.
     ISBN-13: 978-1-59126-051-6
     ISBN-10: 1-59126-051-6
     1. Structural engineering--Examinations, questions, etc. I. Title.

TA638.5.S83 2005
627.1076--dc22
                                                                    2005044623

Table of Contents

About the Author

Christine A. Subasic, PE, is a consulting architectural engineer licensed in North Carolina and Virginia. Ms. Subasic graduated with a Bachelor of Architectural Engineering degree, with honors and high distinction, from Pennsylvania State University. For more than 15 years, she has specialized in structural and masonry design and standards development. Ms. Subasic has design experience in both commercial and residential construction. Her clients include commercial businesses as well as several trade associations.

Ms. Subasic is an active member of ASTM International, an organization committed to the development of construction-industry standards. Ms. Subasic serves on the board of directors for The Masonry Society, where she is active on the Design Practices Committee, Architectural Practices Committee, and Sustainability Subcommittee.

Ms. Subasic is the author of numerous articles and technical publications. She has authored and reviewed chapters in the *Masonry Designers' Guide*, published by The Masonry Society, and co-authored *An Investigation of the Effects of Hurricane Opal on Masonry*. She reviewed and revised the chapters on timber and masonry for the fourth edition of *101 Solved Civil Engineering Problems*. Over the past eight years, her articles on many aspects of masonry have appeared in *Masonry Construction* magazine.

Ms. Subasic is also active at home, with her three boys, at their school, and in her church community.

Preface and Acknowledgments

The Principles and Practice of Engineering examination (PE exam) for civil engineering, prepared by the National Council of Examiners for Engineering and Surveying (NCEES), is developed from sample problems submitted by educators and professional engineers representing consulting, government, and industry. PE exams are designed to test examinees' understanding of both conceptual and practical engineering concepts. Problems from past exams are not available from NCEES or any other source. However, NCEES does identify the general subject areas covered on the exam.

The topics covered in *Six-Minute Solutions for Civil PE Exam Structural Problems* coincide with those subject areas identified by NCEES for the structural engineering depth module of the civil PE exam. Included among these problem topics are Loadings, Analysis, Mechanics of Materials, Materials, Member Design, Failure Analysis, and Design Criteria.

The problems presented in this book are representative of the type and difficulty of problems you will encounter on the PE exam. The book's problems are both conceptual and practical, and they are written to provide varying levels of difficulty. Though you probably won't encounter problems on the exam exactly like those presented here, reviewing these problems and solutions will increase your familiarity with the exam problems' form, content, and solution methods. This preparation will help you considerably during the exam.

Problems and solutions have been carefully prepared and reviewed to ensure that they are appropriate and understandable, and that they were solved correctly. If you find errors or discover an alternative, more efficient way to solve a problem, please bring it to PPI's attention so your suggestions can be incorporated into future editions. You can report errors and keep up with the changes made to this book, as well changes to the exam, by logging on to Professional Publications' website at www.ppi2pass.com/errata.

I would like to dedicate this book to Shawn, my husband and biggest supporter, without whom I could never have worked all the hours necessary to complete this project. I would also like to acknowledge the support of my friends and mentors in the industry, particularly Phillip Samblanet, who always encouraged me in my quest for balance between family and my engineering career, and Maribeth Bradfield, who got me started writing problems in the first place. I am also indebted to Valoree Eikinas, Robert Macia, and all the engineers at Stewart Engineering who "tried out" the problems in this book, and to Thomas H. Miller, PhD, PE, for his work in completing the technical review. Lastly, I thank God for giving me the talents to pursue this.

Christine A. Subasic, PE

Introduction

EXAM FORMAT

The Principles and Practice of Engineering examination (PE exam) in civil engineering is an eight-hour exam divided into a morning and an afternoon session. The morning session is known as the "breadth" exam and the afternoon is known as the "depth" exam.

The morning session includes 40 problems from all of the five civil engineering subdisciplines (environmental, geotechnical, structural, transportation, and water resources), each subdiscipline representing about 20% of the problems. As the "breadth" designation implies, morning session problems are general in nature and wide-ranging in scope.

The afternoon session allows the examinee to select a "depth" exam module from one of the five subdisciplines. The 40 problems included in the afternoon session require more specialized knowledge than do those in the morning session.

All problems from both the morning and afternoon sessions are multiple choice. They include a problem statement with all required defining information, followed by four logical choices. Only one of the four options is correct. Nearly every problem is completely independent of all others, so an incorrect choice on one problem typically will not carry over to subsequent problems.

Topics and the approximate distribution of problems on the afternoon session of the civil structural exam are as follows.

Structural (65%)

- Loadings—dead and live loads, moving loads, wind loads, earthquake loads, repeated loads
- Analysis—determinate, indeterminate, shear diagrams, moment diagrams
- Mechanics of Materials—flexure, shear, torsion, tension and compression, combined stresses, deflection
- Materials—reinforced concrete, prestressed concrete, structural steel, timber, concrete mix design, masonry, composite construction
- Member Design—beams, slabs, columns, reinforced concrete footings, pile foundations, retaining

walls, trusses, braces and connections, shear and bearing walls

- Failure Analysis—buckling, fatigue, failure modes
- Design Criteria—IBC, ACI, PCI, AISC, NDS, AASHTO, ASCE-7

Geotechnical (25%)

- Subsurface Exploration and Sampling—boring log interpretation
- Soil Mechanics Analysis—pressure distribution, lateral earth pressure
- Shallow Foundations—bearing capacity, settlement, proportioning individual/combined footings, mat and raft foundations
- Deep Foundations—axial capacity (single pile/drilled shaft), lateral capacity (single pile/drilled shaft), behavior of pile/drilled shaft groups
- Earth Retaining Structures—gravity walls, cantilever walls, braced and anchored excavations, earth pressure diagrams, stability analysis

Transportation (10%)

- Construction—excavation/embankment, material handling, optimization, scheduling

For further information and tips on how to prepare for the civil environmental engineering PE exam, consult the *Civil Engineering Reference Manual* or Professional Publications' website, www.ppi2pass.com.

THIS BOOK'S ORGANIZATION

Six-Minute Solutions for Civil PE Exam Structural Problems is organized into two sections. The first section, Breadth Problems, presents 20 structural engineering problems of the type that would be expected in the morning part of the civil engineering PE exam. The second section, Depth Problems, presents 80 problems representative of the afternoon part of this exam. The two sections of the book are further subdivided into the topic areas covered by the structural exam.

Most of the problems are quantitative, requiring calculations to arrive at a correct solution. A few are nonquantitative. Some problems will require a little more than six minutes to answer and others a little less. On average, you should expect to complete 80 problems in 480 minutes (eight hours), or spend six minutes per problem.

Six-Minute Solutions for Civil PE Exam Structural Problems does not include problems related directly to transportation and geotechnical engineering, although problems from these subdisciplines will be included in the structural exam. *Six-Minute Solutions for Civil PE Exam Transportation Problems* and *Six-Minute Solutions for Civil PE Exam Geotechnical Problems* provide problems for review in these areas of civil engineering.

HOW TO USE THIS BOOK

In *Six-Minute Solutions for Civil PE Exam Structural Problems*, each problem statement, with its supporting information and answer choices, is presented in the same format as the problems encountered on the PE exam. The solutions are presented in a step-by-step sequence to help you follow the logical development of the correct solution and to provide examples of how you may want to approach your solutions as you take the PE exam.

Each problem includes a hint to provide direction in solving the problem. In addition to the correct solution, you will find an explanation of the faulty solutions leading to the three incorrect answer choices. The incorrect solutions are intended to represent common mistakes made when solving each type of problem. These may be simple mathematical errors, such as failing to square a term in an equation, or more serious errors, such as using the wrong equation.

To optimize your study time and obtain the maximum benefit from the practice problems, consider the following suggestions.

1. Complete an overall review of the problems and identify the subjects that you are least familiar with. Work a few of these problems to assess your general understanding of the subjects and to identify your strengths and weaknesses.

2. Locate and organize relevant resource materials. (See the References section of this book for guidance.) As you work problems, some of these resources will emerge as more useful to you than others. These are what you will want to have on hand when taking the PE exam.

3. Work the problems in one subject area at a time, starting with the subject areas that you have the most difficulty with.

4. When possible, work problems without utilizing the hint. Always attempt your own solution before looking at the solutions provided in the book. Use the solutions to check your work or to provide guidance in finding solutions to the more difficult problems. Use the incorrect solutions to help identify pitfalls and to develop strategies to avoid them.

5. Use each subject area's solutions as a guide to understanding general problem-solving approaches. Although problems identical to those presented in *Six-Minute Solutions for Civil PE Exam Structural Problems* will not be encountered on the PE exam, the approach to solving problems will be the same.

Solutions presented for each example problem may represent only one of several methods for obtaining a correct answer. Although most of these problems have unique solutions, alternative problem-solving methods may produce a different, but nonetheless appropriate, answer.

References

The minimum recommended library for the civil exam consists of PPI's *Civil Engineering Reference Manual*. You may also find the following references helpful in completing some of the problems in *Six-Minute Solutions for Civil PE Exam Structural Problems*.

American Association of State Highway and Transportation Officials (AASHTO). *Standard Specification for Highway Bridges*.

American Concrete Institute. *Building Code Requirements for Structural Concrete* (ACI 318).

American Forest & Paper Association/American Wood Council. *National Design Specification for Wood Construction* (NDS), with supplement.

American Institute of Steel Construction (AISC). *Manual of Steel Construction: Allowable Stress Design* (ASD).

_____. *Manual of Steel Construction: Load and Resistance Factor Design* (LRFD).

American Institute of Timber Construction. *Standard Specification for Structural Glued Laminated Timber of Softwood Species, Design Requirements* (AITC 117).

American Society of Civil Engineers. *Minimum Design Standards for Buildings and Other Structures* (ASCE 7).

ASTM International. *Standard Specification for Hollow Brick (Hollow Masonry Units Made from Clay or Shale)* (C652). ASTM International.

Bowles, Joseph E. *Foundation Analysis and Design*. New York: McGraw-Hill.

International Code Council. *International Building Code* (IBC).

Masonry Society, The. *Masonry Designers' Guide* (MGD-4), 4th ed. Boulder, CO: The Masonry Society.

Masonry Standards Joint Committee (MSJC). *Building Code Requirements for Masonry Structures* (ACI 530/ASCE 5/TMS 402).

_____. *Specification for Masonry Structures* (ACI 530.1/ASCE 6/TMS 602).

McCormac, James K. *Structural Analysis*, 4th ed. New York: Harper & Row.

National Concrete Masonry Association (NCMA). *Concrete Masonry Wall Weights* (TEK 14-13A).

Nilson, Arthur H. and George Winter. *Design of Concrete Structures*. New York: McGraw-Hill.

Codes Used to Prepare This Book

At the time of printing, the codes used to write these problems corresponded with those announced by the NCEES as being the basis of the structural portion of the Civil PE Exam. However, the codes used on the exam are subject to change as NCEES sees fit. For a current list of the codes used by the NCEES, refer to PPI's website (www.ppi2pass.com).

American Association of State Highway and Transportation Officials (AASHTO). *Standard Specification for Highway Bridges*, 17th ed. 2002.

American Concrete Institute. *Building Code Requirements for Structural Concrete* (ACI 318). 2002.

American Forest & Paper Association/American Wood Council. *National Design Specification for Wood Construction* (NDS), with supplement. 2001.

American Institute of Steel Construction (AISC). *Manual of Steel Construction: Allowable Stress Design* (ASD), 9th ed. 1989.

_____. *Manual of Steel Construction: Load and Resistance Factor Design* (LRFD), 3rd ed. 2001.

American Society of Civil Engineers. *Minimum Design Standards for Buildings and Other Structures* (ASCE 7). 2002.

International Code Council. *International Building Code* (IBC). 2003.

Masonry Standards Joint Committee (MSJC). *Building Code Requirements for Masonry Structures* (ACI 530/ASCE 5/TMS 402). 2002.

_____. *Specification for Masonry Structures* (ACI 530.1/ASCE 6/TMS 602). 2002.

Nomenclature

Symbol	Description	US units	SI units
a	dimension	ft	m
A	area	in^2, ft^2	mm^2, m^2
A	cross-sectional area	in^2, ft^2	mm^2, m^2
A_{cp}	area enclosed by outside perimeter of concrete	in^2	mm^2
A_t	cross-sectional area of torsion reinforcement	in^2	mm^2
b	dimension	in	mm
b	dimension from web to centerline of bolt hole in hanger connection	in	mm
b	width	in	mm
b	width of footing	ft	m
B	allowable tension per bolt	kips	N
B	width of column base plate in direction of column flange	in	mm
B	width of footing	ft	m
b_a	tension per bolt	kips, lbf	N
B_a	allowable tension per bolt	kips	N
b_e	effective width of slab	in	mm
b_v	shear per bolt	kips, lbf	N
B_v	allowable shear per bolt	lbf	N
c	neutral axis depth	in	mm
c	undrained shear strength (cohesion)	lbf/ft^2	Pa
C	correction factor	various	various
c_A	adhesion	lbf/ft^2	Pa
C_1	moment coefficient	–	–
d	beam depth	in	mm
d	beam depth (to tension steel centroid)	in	mm
d	bolt diameter	in	mm
d	diameter	in, ft	mm, m
d	effective depth	in	mm
d'	diameter of bolt hole	in	mm
D	dead load	kips, lbf, lbf/ft^2	N, Pa
D	depth	ft, in	m
D	diameter	ft	m
d_b	nominal diameter of reinforcing bar	in	mm
DF	distribution factor	–	–
e	eccentricity	in, ft	mm, m
E	earthquake load	kips	N
E	modulus of elasticity	kips/in^2, lbf/ft^2, lbf/in^2	Pa
E'	allowable modulus of elasticity	kips/in^2, lbf/ft^2, lbf/in^2	Pa
f	stress	lbf/in^2	Pa
F	allowable stress	kips/in^2	Pa
F	factor of safety	–	–
F	force	lbf, kips	N
F	strength	kips/in^2	Pa
F'	reduced allowable stress	kips/in^2	Pa
f_a	compressive stress in masonry due to axial load alone	lbf/in^2	Pa
F_a	allowable compressive stress due to axial load alone	lbf/in^2	Pa
f_b	bending stress	lbf/in^2	Pa
f_b	stress in masonry due to flexure alone	lbf/in^2	Pa
F_b	allowable bending stress	lbf/in^2	Pa
F_b	allowable compressive stress due to flexure alone	lbf/in^2	Pa
f'_c	compressive strength of concrete	lbf/in^2	Pa
f'_m	compressive strength of masonry	lbf/in^2	Pa
F_p	allowable bearing pressure	lbf/in^2	Pa
f_r	modulus of rupture	lbf/in^2	Pa
f_s	shear stress	lbf/in^2	Pa
f_s	stress in steel reinforcement	lbf/in^2	Pa
F_s	allowable tensile stress in steel reinforcement	lbf/in^2	Pa
f_t	tensile stress	lbf/in^2	Pa
F_t	allowable tensile stress	lbf/in^2	Pa
f_v	shear stress in masonry	lbf/in^2	Pa
F_v	allowable shear stress in masonry	lbf/in^2	Pa
f_y	yield stress	lbf/in^2	Pa
F_y	yield strength of steel	lbf/in^2	Pa
FEM	fixed-end moment	ft-kips	N·m

Symbol	Description	US	SI
g	lateral gage spacing of adjacent holes	in	mm
g	ratio of distance between tension steel and compression steel to overall column depth	–	–
G	gust factor	–	–
h	distance	ft	m
h	effective height	in	mm
h	height	in, ft	mm, m
h	overall thickness	in, ft	mm, m
h'	modified overall thickness	in, ft	mm, m
H	distance from soil surface to footing base	ft	m
H_s	length of stud connector	in	mm
I	moment of inertia	in^4	mm^4
I_p	polar moment of inertia	in^4	mm^4
j	ratio of distance between centroid of flexural compressive forces and centroid of tensile forces to depth	–	–
k	coefficient	–	–
k	coefficient of lateral earth pressure	–	–
k	effective length factor	–	–
k	stiffness	lbf/ft	N/m
K	coefficient	–	–
K	effective length factor	–	–
K	relative stiffness	ft^{-1}	m^{-1}
l	distance between points of lateral support of compression member in a given plane	in, ft	mm, m
l	length	in, ft	mm, m
l	span length	in, ft	mm, m
L	length	in, ft	m
L	length of footing	ft	m
L	live load	kips, lbf, lbf/ft^2	N, Pa
L	span length	in, ft	mm, m
l_c	vertical distance between supports	in	mm
l_d	development length of reinforcement	in	mm
l_h	distance from center of bolt hole to beam end	in	mm
l_n	clear span length	in, ft	mm, m
l_u	unsupported length of a compression member	in	m
l_v	distance from center of bolt hole to edge of web	in	mm
M	moment	ft-kips, in-lbf, ft-lbf	N·mm, N·m
M_a	maximum moment	in-lbf	N·m
M_m	resisting moment assuming masonry governs	in-lbf	N·m
M_o	total factored static moment	in-lbf	N·m
M_R	beam resisting moment	ft-kips	N·mm
M_s	resisting moment assuming steel governs	in-lbf	N·m
M_t	torsional moment	ft-kips	N·mm
n	modular ratio	–	–
n	quantity (number of)	–	–
N	bearing capacity factor	–	–
N	length of column base plate in direction of column depth	in	mm
N_r	number of studs in one rib	–	–
p	perimeter	in, ft	mm, m
p	pressure	lbf/ft^2	Pa
P	axial load	kips	N
P	load	kips	N
P	prestress force in a tendon	lbf, kips	N
P_a	allowable compressive force in reinforced masonry due to axial load	kips	N
p_{cp}	outside perimeter of concrete	in	mm
P_e	Euler buckling load	kips	N
P_{nw}	nominal axial strength of a wall	kips	N
P_o	maximum allowable axial load for zero eccentricity	kips	N
q	allowable horizontal shear per shear connector	kips	N
q	soil pressure under footing	lbf/ft^2	Pa
q	uniform surcharge	lbf/ft, lbf/ft^2	N/m, Pa
Q	bearing capacity	lbf/ft^2	Pa
Q	nominal load effect	various	various
Q	statical moment	in^3	mm^3
q_c	tip resistance	lbf/ft^2	Pa
q_s	skin friction resistance	lbf/ft^2	Pa
r	radius of gyration	in, ft	mm, m
r	rigidity	–	–
R	allowable load per bolt	kips	N
R	concentrated load	lbf, kips	N
R	reaction	lbf, kips	N
R	resultant force	lbf	N
RF	reduction factor	–	–
s	spacing	in	mm
S	force	lbf	N
S	section modulus	in^3, ft^3	mm^3, m^3
S	snow load	kips	N
S_{SD}	design earthquake spectral response accelerations at short period	–	–

Symbol	Description	US units	SI units
S_{MS}	maximum considered earthquake spectral response accelerations at short period	–	–
t	nominal weld size	in	mm
t	slab thickness	in	mm
t	thickness	in, ft	mm, m
t	wall thickness	in, ft	mm, m
T	period	sec	s
T	temperature	°F, °R	°C, K
T	tension force	kips, lbf	N
T	torsional moment (torque)	ft-lbf	N·m
t_e	effective throat thickness of a weld	in	mm
T_u	factored torsional moment	ft-lbf	N·m
u	unit force	–	–
U	ultimate strength required to resist factored loads	lbf	N
v	shear stress	lbf/in²	Pa
v	wind speed	mph	kph
V	design shear force	lbf	N
V	shear	kips, lbf	N
V	shear strength	lbf, lbf/in²	N, Pa
v_c	allowable concrete shear stress	lbf/in²	Pa
V_c	allowable concrete shear strength	lbf	N
v_u	shear stress due to factored loads	lbf/in²	Pa
V_u	factored shear force	lbf	N
w	distributed load	lbf/ft	N/m
w	uniformly distributed load	lbf/ft, lbf/ft²	N/m, Pa
w	weld size	in	mm
w	tributary width	in, ft	mm, m
w'	tributary width	in, ft	mm, m
W	effective seismic weight	lbf, kips	N
W	weight	lbf	N
W	wind load	lbf	N
x	distance	in, ft	mm, m
x	location	in	mm
x	x-coordinate of position	ft	m
$\overline{x}$	distance to center of rigidity in the x-direction	in, ft	mm, m
$\overline{y}$	distance from centroidal axis to the centroid of the area	in	mm
$\overline{y}$	distance to center of rigidity in the y-direction	in, ft	mm, m
y	location	in	m
y	y-coordinate of position	ft	m
y_c	distance from top of the section to the centroid of the section	in	mm
y_t	distance from centroid of the section to the extreme fiber in tension	in	mm

Symbols

Symbol	Description	US units	SI units
α	coefficient of linear thermal expansion	°F⁻¹	°C⁻¹
α	ratio of flexural stiffness of beams in comparison to slab	–	–
β	column strength factor	–	–
β	ratio of clear spans in long-to-short directions of a two-way slab	–	–
β	ratio of long side to short side of a footing	–	–
γ	ratio of the distance between bars on opposite faces of a column to the overall column dimension, both measured in the direction of bending	–	–
γ	specific weight (unit weight)	lbf/ft³	N/m³
Δ	deflection	in	mm
θ	angle	deg, rad	deg, rad
λ	distance from centroid of compressed area to extreme compression fiber	in	mm
λ	height and exposure adjustment factor	–	–
ρ	reinforcement ratio	–	–
ρ_g	longitudinal reinforcement ratio	–	–
ρ_h	ratio of horizontal shear reinforcement to gross concrete area		
ρ_n	ratio of vertical shear reinforcement area to gross concrete area for a shear wall	–	–
σ	normal stress	lbf/ft²	Pa
τ	shear stress	lbf/ft²	Pa
ϕ	strength reduction factor	–	–
Ψ	relative stiffness parameter	–	–

Subscripts

γ	density
0	initial
a	active, allowable, or due to axial loading
A	adhesion
b	beam, bending, or bolt
bm	beam
bot	bottom
BS	block shear
c	cohesive, column, concrete, or curvature
cr	cracked, cracking

cs   critical section
d directionality or penetration depth
D dead load or duration
e effective or exposure
E Euler
eg   end grain
f flange, flat, force, form, or skin friction
F size
fs   face shell
fu   flat use
g gross or grout
h horizontal
i initial, inside, or ith member
j jth member
k kern or kth member
l longitudinal
L beam stability, left, or live load
m masonry
M wet service
max   maximum
min   minimum
n nail, net, or nominal
ns   nonsway frame
o centroidal or initial
OT   overturning
p bearing, passive, pile tip, plate, or prestressed
prov   provided

q surcharge
r rafter, reduced, repetitive, resultant, rib, roof, or rupture
R resistance, resisting, resistive, resultant, or right
req   required
s shear, side friction, simplified, skin, snow, spiral, or steel reinforcement
sat   saturated
sd   short direction
SL   sliding
sp   single pile
sr   stress range
t temperature, tensile, tension, topography, torsion, or tributary
th   thermal
tr   transformed
u ultimate (factored), ultimate tensile, unbraced, or untreated
v shear or vertical
V volume
w wall, web, weld, or wind
x at a distance x, in x-direction, or strong axis
y in y-direction, weak axis, or yield
z at height z

Breadth Problems

LOADINGS
PROBLEM 1

A steel beam supports a 6 in lightweight concrete masonry (CMU) wall around a mechanical room. The CMU has a unit weight, γ, of 85 lbf/ft^3. If the wall is 8 ft high and 10 ft long, the load on the beam is most nearly

(A) 340 lbf/ft
(B) 680 lbf/ft
(C) 3400 lbf/ft
(D) 4100 lbf/ft

Hint: The load on the beam is uniformly distributed.

PROBLEM 2

A two-story office building has a 60 ft by 100 ft rectangular floor plan. Steel columns spaced 20 ft apart carry the following loads from the roof and second floor. Live load reductions are not permitted.

roof dead load	15 lbf/ft^2
roof live load	20 lbf/ft^2
floor dead load	15 lbf/ft^2
floor live load	60 lbf/ft^2

What is the total load on an interior first-floor column?

(A) 2.2 kips
(B) 14 kips
(C) 44 kips
(D) 66 kips

Hint: Determine the tributary area for an interior column.

ANALYSIS
PROBLEM 3

Analyze the truss shown.

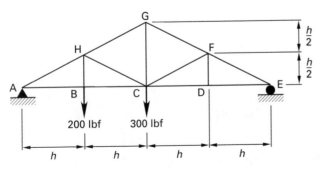

The force in member AH is most nearly

(A) 300 lbf (compression)
(B) 600 lbf (compression)
(C) 670 lbf (compression)
(D) 670 lbf (tension)

Hint: Solve by using the method of joints.

PROBLEM 4

A single-story flex warehouse building is constructed of 8 in concrete masonry walls that have been grouted solid. One 20 ft long wall contains a 4 ft opening centered in the wall as shown. A steel lintel made from two 5 in by 3^1/$_2$ in by ^{1}/$_2$ in angles spans the opening. A 700 lbf concentrated load from a roof truss is centered over the opening. The unit weight of the wall is 150 lbf/ft^3.

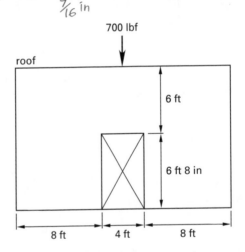

What is the design moment on the lintel?

(A) 300 ft-lbf
(B) 400 ft-lbf
(C) 1000 ft-lbf
(D) 1200 ft-lbf

Hint: The wall described will exhibit arching action over the opening.

PROBLEM 5

A 20 ft beam carries a load that decreases linearly from a maximum of 350 lbf/ft at its left support to 0 lbf/ft at the midspan of the beam. Measured from the left support, the point at which the shear is zero is most nearly

(A) 0.87 ft
(B) 4.1 ft
(C) 5.9 ft
(D) 14 ft

Hint: Begin by drawing the free-body diagram.

PROBLEM 6

A concrete column and footing carries the loads shown. The footing is 6 ft wide.

column dead load	80 kips
column live load	100 kips
moment due to wind	240 ft-kips
compressive strength of concrete	3000 lbf/in^2
maximum allowable soil pressure	3000 lbf/ft^2

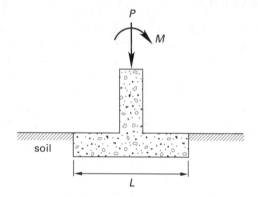

What is the minimum footing length, L, required for the entire footing to be considered effective in carrying these loads?

(A) 7.7 ft
(B) 8.0 ft
(C) 8.7 ft
(D) 10 ft

Hint: For the entire footing to be effective, avoid tensile stress in the soil.

MECHANICS OF MATERIALS
PROBLEM 7

A three-span continuous steel beam carries a uniform dead load of 500 lbf/ft and a uniform live load of 1000 lbf/ft. It is an A36 W16 × 31 beam, and each span is 20 ft.

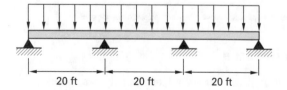

What is the maximum deflection of the first span?

(A) 0.21 in
(B) ~~0.26 in~~ 0.27 in
(C) 0.33 in
(D) 0.50 in

Hint: Use the beam loading diagrams found in the AISC ASD manual.

PROBLEM 8

The cantilevered beam shown has a varying moment of inertia. The moment of inertia for the first 15 ft, I_1, is 2000 in^4. The second moment of inertia, I_2, is 1000 in^4. The modulus of elasticity of the beam is 29×10^6 lbf/in^2.

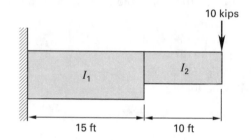

The deflection at the free end of the beam is most nearly

(A) 0.0017 in
(B) 1.7 in
(C) 3.0 in
(D) 3.1 in

Hint: Use the moment-area method to find the deflection of the beam.

PROBLEM 9

The T-shaped beam shown carries a shear load of 100 lbf.

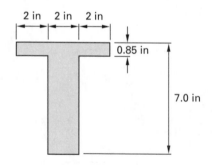

The shear stress at the centroid of the cross section is most nearly

(A) 6.5 lbf/in^2
(B) 8.1 lbf/in^2
(C) 10 lbf/in^2
(D) 22 lbf/in^2

Hint: First find the centroid of the cross section.

PROBLEM 10

A column with the cross section shown carries an axial load of 560 lbf applied at the centroid of the section.

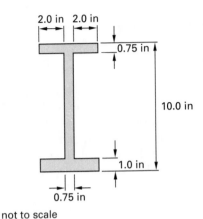

not to scale

The compressive stress for the cross section is

- (A) 15 lbf/in^2
- (B) 35 lbf/in^2
- (C) 39 lbf/in^2
- (D) 42 lbf/in^2

Hint: An axial load applied at the centroid of the section creates uniformly distributed stress.

MATERIALS

PROBLEM 11

A concrete slab designed for a parking garage constructed in Chicago uses normal-weight concrete with a compressive strength of 3000 lbf/in^2. If the maximum aggregate size is 1 in, the total percentage of air content of the concrete mix should be

- (A) 0.06%
- (B) 0.45%
- (C) 4.5%
- (D) 6%

Hint: The parking garage in Chicago is exposed to freezing and thawing conditions.

PROBLEM 12

Which of the following statements are true regarding reinforced concrete slabs designed using ACI 318?

I.   The Direct Design Method applies only to two-way slab systems.
II. The Direct Design Method can be used for slabs with five continuous spans in one direction and two continuous spans in the other direction.
III. The number and length of spans is not restricted using the Equivalent Frame Method of slab design.

IV. A slab with a dead load of 50 lbf/ft^2 and a live load of 150 lbf/ft^2 cannot be designed using the Direct Design Method, but can be designed using the Equivalent Frame Method.

- (A) I and II
- (B) I and III
- (C) I, III, and IV
- (D) I, II, III, and IV

Hint: Refer to ACI 318 Ch. 13.

PROBLEM 13

Which of the following statements is/are true regarding the shear capacity of a fillet weld?

I.   It is limited by the effective throat thickness of the weld.
II. It is directly proportional to the length of the weld.
III. It is determined by the direction of the weld.
IV. It is limited by the tensile strength of the weld metal.

- (A) III
- (B) I and II
- (C) I, II, and III
- (D) I, II, and IV

Hint: Refer to the AISC ASD specification, Secs. J2.2–J2.5.

PROBLEM 14

A W12 × 72 structural steel column on an 8 ft by 8 ft concrete-spread footing supports a load of 420 kips.

compressive strength of concrete	3000 lbf/in^2
yield stress of steel reinforcement	36 kips/in^2

Using the AISC ASD manual, if the steel base plate size is limited to 14 in by 16 in, the thickness of the base plate must be most nearly

- (A) 1.00 in
- (B) 1.39 in
- (C) 1.45 in
- (D) 1.76 in

Hint: Start by determining the allowable bearing pressure on the concrete.

PROBLEM 15

Using the NDS, which of the following statements must be true?

I.   The temperature factor, C_t, applies to members subjected to extremely cold temperatures.
II. The volume factor, C_V, applies only to glued laminated timber and structural composite lumber bending members.

III. The bending design allowable stress, F_b, for a floor framed with 1×6 sawn lumber joists must be multiplied by the repetitive member factor.

IV. The load duration factor, C_D, does not apply to the modulus of elasticity values.

    (A) I and II
    (B) II and III
    (C) II and IV
    (D) III and IV

Hint: Refer to NDS Sec. 2.3 for adjustment of design values.

MEMBER DESIGN

PROBLEM 16

The three-story, 4 in brick veneer ($40 \ \text{lbf/ft}^2$) office building shown is framed with Grade 36 structural steel. Assume simple connections and full lateral support of the beam. The floor dead load is $60 \ \text{lbf/ft}^2$, and the floor live load is $40 \ \text{lbf/ft}^2$. Total deflection should not exceed $L/600$ or 0.3 in. Design using the AISC ASD manual.

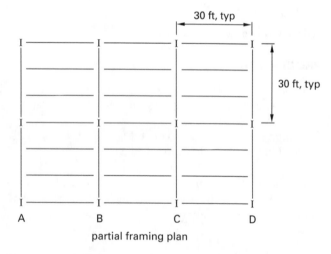

partial framing plan

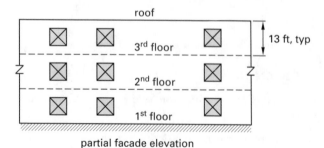

partial facade elevation

What is the minimum size of third-floor spandrel beam BC for a typical 30 ft by 30 ft bay?

    (A) W14 $\times$ 22
    (B) W16 $\times$ 40
    (C) W24 $\times$ 55
    (D) W24 $\times$ 84

Hint: The weight of the facade is typically supported by spandrel beams at each floor.

PROBLEM 17

A steel column supports a concentric load of 525 kips. The effective length with respect to the major axis is 32 ft. The effective length with respect to the minor axis is 18 ft. Using allowable stress design, what is the lightest W shape that can be used for the column if Grade 50 steel is used and the column depth cannot exceed 12 in (nominal)?

    (A) W12 $\times$ 87
    (B) W12 $\times$ 120
    (C) W12 $\times$ 170
    (D) W12 $\times$ 190

Hint: Use the columns tables in Part 3 (Column Design) of the AISC ASD manual.

PROBLEM 18

A circular spiral concrete column supports a 300 kip dead load and a 350 kip live load. The concrete compressive strength is $4000 \ \text{lbf/in}^2$, and the yield stress of the reinforcement is $60{,}000 \ \text{lbf/in}^2$. If the maximum reinforcement is used, the cross-sectional area of the column is most nearly

    (A) ~~130 in²~~ $140 \ \text{in}^2$
    (B) $150 \ \text{in}^2$
    (C) $200 \ \text{in}^2$
    (D) ~~230 in²~~ $210 \ \text{in}^2$

Hint: Refer to ACI 318 Sec. 10.3.6.

PROBLEM 19

A 6.5 ft by 6.5 ft concrete spread footing supports a centrally located 12 in by 12 in concrete column. The dead load on the column, including the column weight, is 50 kips. The live load is 75 kips. The soil report indicates an allowable soil pressure of $4000 \ \text{lbf/ft}^2$. Disregard the weight of the soil. If the concrete compressive strength, f'_c, is $3000 \ \text{lbf/in}^2$ and no. 5 bars are used each way in the footing, the minimum thickness of the footing is most nearly

    (A) 6.1 in
    (B) 11 in
    (C) 12 in
    (D) 15 in

Hint: Two-way shear action controls for centrally loaded square footings.

PROBLEM 20

Lightweight concrete (100 lbf/ft^3) is used for the one-way slab shown.

compressive strength of concrete	3000 lbf/in^2
yield stress of reinforcement steel	60,000 lbf/in^2

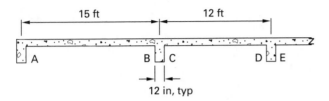

What is the minimum thickness for slab AB?

(A) 7.5 in
(B) ~~8.0 in~~ 8.9 in
(C) 8.6 in
(D) 10 in

Hint: Minimum thicknesses of one-way slabs not supporting partitions or other elements likely to be damaged by deflection can be determined using ACI 318 Table 9.5(a).

Depth Problems

LOADINGS

PROBLEM 21

A fire station located in Utah is built with concrete masonry bearing walls and ordinary reinforced concrete masonry shear walls. The walls of the fire station are 20 ft tall. The maximum considered earthquake spectral response acceleration for the short period is 2.12. Using the simplified analysis procedure for seismic design found in the IBC, the seismic base shear is

(A) $0.38W$
(B) $0.57W$
(C) $0.68W$
(D) $1.0W$

Hint: Refer to IBC Sec. 1617.5 for simplified seismic analysis.

PROBLEM 22

Which of the following seismic design statements are true?

I.   The seismic base shear of a building increases as the building dead load increases.
II.   In areas of high seismic activity, it is best to design buildings with an irregular floor plan to better break up the seismic load.
III. According to the IBC, flat roof snow loads under 30 lbf/ft^2 need not be included in the effective seismic weight of the structure, W.
IV.   Buildings with a soft first story and heavy roofs performed well during the Northridge earthquake in 1994.

(A) I and II
(B) I and III
(C) II and IV
(D) I, II, and III

Hint: Chapter 16 of the IBC provides information on seismic design.

PROBLEM 23

A 2.5 in diameter mild steel bar is fixed at each end by a steel plate. The modulus of elasticity of the steel is 29×10^6 lbf/in^2. At 8:00 a.m., the bar has a temperature of 60°F and experiences zero stress. At 3:30 p.m., the temperature of the bar is 125°F. Ignore gravity loads. The axial load in the bar at 3:30 p.m. is most nearly

(A) 60 kips
(B) 120 kips
(C) 240 kips
(D) 6000 kips

Hint: Part 6 of the AISC ASD manual contains information on section properties and thermal expansion properties of steel.

PROBLEM 24

A building is constructed with 9 ft high shear walls as shown.

wall A: 16 in thick, 60 ft long
wall B: 16 in thick, 36 ft long
wall C: 12 in thick, 40 ft long
wall D: 12 in thick, 36 ft long
wall E: 16 in thick, 36 ft long

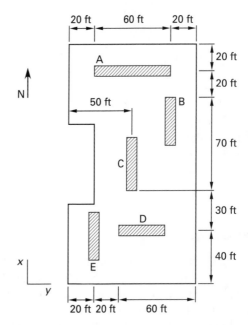

The center of rigidity of the building is located at a point that is

- (A) 115 ft from the South edge of the building and 50.0 ft from the East edge of the building.
- (B) 115 ft from the South edge of the building and 54.9 ft from the West edge of the building.
- (C) 123 ft from the North edge of the building and 50.0 ft from the East edge of the building.
- (D) 123 ft from the South edge of the building and 50.0 ft from the West edge of the building.

Hint: The center of rigidity can be based on the relative areas of the walls.

PROBLEM 25

A one-story steel-framed building has earth bermed on one side. The total resultant load from the earth pressure is 600 kips and is resisted entirely by the three shear walls as shown. Wall A is 12 in thick. Walls B and C are 14 in thick. The roof is framed with steel joists and corrugated steel deck (no concrete).

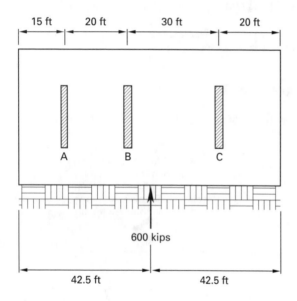

What is the shear load on wall A?

- (A) 124 kips
- (B) 176 kips
- (C) 180 kips
- (D) 247 kips

Hint: A steel joist and corrugated deck roof system is typically considered a flexible diaphragm unless a concrete slab is poured on the roof deck.

PROBLEM 26

A two-story office building located in a South Dakota office park is 40 ft by 60 ft in plan. The asphalt-shingle hip roof has a 6:12 pitch and is supported by trusses spanning the short direction of the building. The attic is vented and insulated with R-30 insulation. The local

building code official requires design using the IBC. If the ground snow load is 40 lbf/ft², what is the maximum leeward snow load?

- (A) 31 lbf/ft²
- (B) 32 lbf/ft²
- (C) 46 lbf/ft²
- (D) 51 lbf/ft²

Hint: The leeward side of the roof is the side away from the wind.

PROBLEM 27

A building is designed with walls that act in shear only to resist the wind load as shown. The north wind load is 200 lbf/ft. All of the walls are 12 in thick. The center of rigidity (c.r.) of the building is located 98.3 ft from the West facade and 40 ft from the South facade.

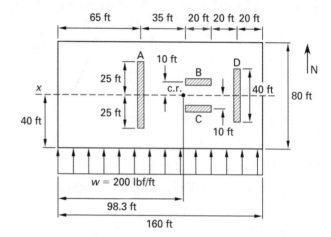

What is the shear load on wall A?

- (A) 6.6 kips
- (B) 12 kips
- (C) 18 kips
- (D) 25 kips

Hint: Wind load is distributed to shear walls according to the relative rigidity of the walls.

ANALYSIS

PROBLEM 28

A rectangular combined footing is used near a property line to support two 12 in square concrete columns as shown. The allowable soil pressure is 2000 lbf/ft². The compressive strength of concrete is 3000 lbf/in².

Column 1:

dead load	30 kips
live load	60 kips
moment due to dead load	30 ft-kips
moment due to live load	40 ft-kips

Column 2:

dead load	60 kips
live load	60 kips
moment due to dead load	30 ft-kips
moment due to live load	40 ft-kips

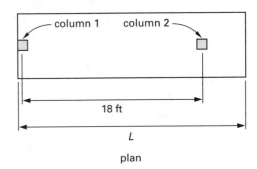

plan

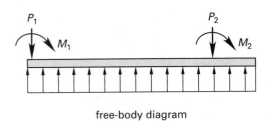

free-body diagram

What minimum footing length, L, will create uniform soil pressure?

(A) 21.6 ft
(B) 21.8 ft
(C) 22.6 ft
(D) 23.0 ft

Hint: For the soil pressure to be uniform, the resultant, R, must be at the centroid of the footing area.

PROBLEM 29

A pin-connected truss is used to support a sign as shown. The product of the area and the modulus of elasticity equals 3 kips for all members except AD and BC, for which the product of the area and the modulus of elasticity equals 5 kips.

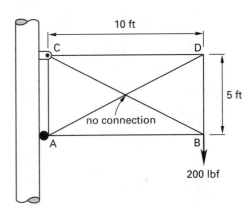

What is the force in member BC?

(A) −209 lbf (compression)
(B) 234 lbf (tension)
(C) 238 lbf (tension)
(D) 447 lbf (tension)

Hint: Assess whether the truss is determinate.

PROBLEM 30

A beam with a constant cross section and constant modulus of elasticity is loaded as shown. The relative stiffness of section AB is 0.286/ft and of section BC is 0.190/ft.

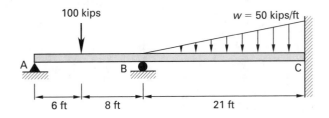

Using moment distribution, the moment at the fixed end is most nearly

(A) 1050 ft-kips, counterclockwise
(B) 1220 ft-kips, clockwise
(C) 1220 ft-kips, counterclockwise
(D) 1270 ft-kips, clockwise

Hint: See a reference containing fixed-end moments.

PROBLEM 31

A moment distribution analysis of a beam determined the moments at the beam supports as shown.

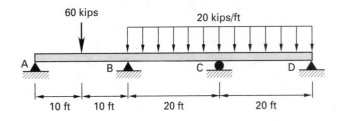

	A	B		C		D
DF	1.0	0.5	0.5	0.5	0.5	1.0
FEM	−150.0	+150.0	−666.7	+666.7	−666.7	+666.7

Moment distribution gives the following results.

	A	B	C	D
M	0	+386.8 −386.8	+903.2 −903.2	0

The reaction at support B is most nearly

- (A) 50 kips
- (B) 120 kips
- (C) 170 kips
- (D) 220 kips

Hint: Use free-body diagrams to determine the shear at the supports.

PROBLEM 32

A warehouse facility is earth-bermed to the midpoint of the wall on one side as shown.

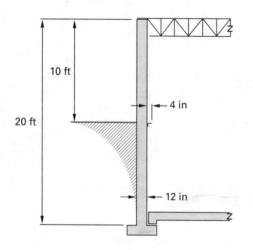

The exterior walls are 12 in reinforced concrete with a compressive strength of 3000 lbf/in². Steel joists framing the roof bear on the full width of the top of the concrete wall. An L6 × 4 × ½ (LLV) angle bolted to the concrete wall is used to secure a storage rack and is designed to carry a load of 75 lbf/ft. The first floor is a concrete slab on grade. Assume the wall is simply supported and disregard the effects of wind. The gravity loads on the wall due to the roof are uniformly distributed and are as follows.

| dead load from roof | 800 lbf/ft |
| live load from roof | 1000 lbf/ft |

If the soil specific weight is 45 lbf/ft³, the maximum unfactored wall bending moment is most nearly

- (A) 4260 ft-lbf/ft
- (B) 4350 ft-lbf/ft
- (C) 4750 ft-lbf/ft
- (D) 4820 ft-lbf/ft

Hint: To determine the maximum moment, consider the eccentricity of the loads.

PROBLEM 33

For the one-way beam shown, use the moment coefficients found in ACI 318 to determine the largest magnitude moment and its location. Support A is a beam, supports BC and DE are built-in columns, and support F is a simple support. The beam carries a uniform factored load of 600 lbf/ft, and the live load is 1.5 times the dead load.

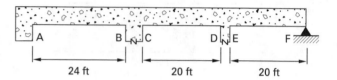

The critical moment is most nearly

- (A) −38,000 ft-lbf at point B
- (B) −35,000 ft-lbf at point B
- (C) −29,000 ft-lbf at point B
- (D) −24,000 ft-lbf at point E

Hint: Refer to ACI 318 Sec. 8.3.

PROBLEM 34

A flat-plate concrete slab is supported by 12 in square concrete columns. The floor dead load is 100 lbf/ft². The floor live load is 30 lbf/ft². The design moments of the interior slab panel shown are to be determined using the Direct Design Method.

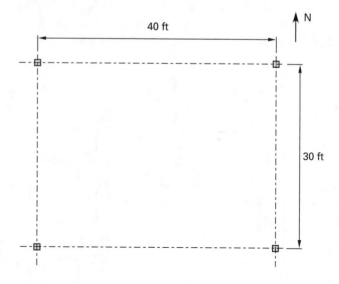

The midspan moment of the middle strip in the East/West direction is most nearly

(A) 88 ft-kips   77 ft-kips
(B) 110 ft-kips   99 "
(C) 150 ft-kips   130 "
(D) 160 ft-kips   140 "

Hint: Refer to ACI 318 Ch. 13.

MECHANICS OF MATERIALS

PROBLEM 35

A 12 in unreinforced concrete masonry wall supports joists spaced 12 in on center. The reaction from each joist is 700 lbf. The wall is grouted solid.

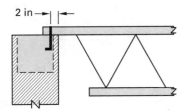

The stress on the wall due to the joists is most nearly

(A) 5.0 lbf/in^2 (compression)
(B) 5.0 lbf/in^2 (tension)
(C) 5.0 lbf/in^2 (compression), 10 lbf/in^2 (tension)
(D) 15 lbf/in^2 (compression), 5.0 lbf/in^2 (tension)

Hint: Determine the eccentricity of the load.

PROBLEM 36

The beam shown has a modulus of elasticity of 1.2×10^6 lbf/in^2 and a moment of inertia of 240 in^4.

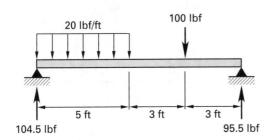

Using the conjugate beam method, the deflection at a point 5 ft from the left support of the beam is most nearly

(A) 1.1×10^{-5} in
(B) 1.5×10^{-4} in
(C) 2.2×10^{-2} in
(D) 3.1×10^{-2} in

Hint: Begin by drawing the M/EI diagram for the beam.

PROBLEM 37 (including self weight)

A simply supported reinforced concrete beam carries a uniform dead load of 800 lbf/ft and a uniform live load of 1200 lbf/ft. The beam span length is 24 ft. The beam is constructed of normal-weight concrete with a compressive strength of 6000 lbf/in^2.

beam width	16 in
beam height	30 in
beam depth to reinforcement	24 in
area of reinforcement steel	7.90 in^2

The maximum immediate deflection is most nearly

(A) 0.094 in
(B) 0.13 in
(C) ~~0.22 in~~ 0.20 in
(D) 0.29 in

Hint: Immediate deflection does not include creep and shrinkage effects.

PROBLEM 38

A 20 ft W12 × 65 steel column supports a 30 kip concentric roof load at the top of the column and a 50 kip eccentric load at mid-height as shown. The column is pinned at both ends in each direction. Lateral support is provided in the weak direction at mid-height.

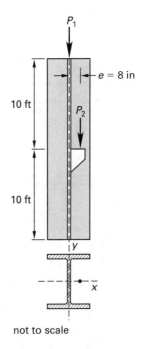

not to scale

Using the AISC ASD manual, if the yield stress is 36 kips/in^2 and the allowable bending stress has been determined to be 22 kips/in², the critical combined stress ratio for axial load and flexure is most nearly

   (A) 0.550
   (B) 0.555
   (C) 0.651
   (D) 1.27

Hint: Examine buckling in each direction.

PROBLEM 39

The beam shown carries a uniform eccentric load of 20 kips/ft. The span is 20 ft and the beam is fully restrained at one end.

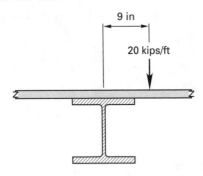

What is the maximum torsional moment on the beam?

   (A) 150 ft-kips
   (B) 300 ft-kips
   (C) 670 ft-kips
   (D) 1000 ft-kips

Hint: Torsion is the product of a force and an eccentricity.

PROBLEM 40

The 10 ft long channel beam shown is made from $\frac{1}{2}$ in thick flat plates welded together and is simply supported at both ends. A load is applied at the midspan of the beam.

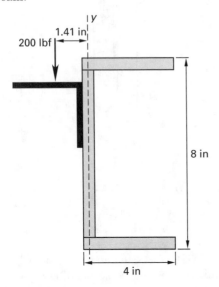

The load on the beam is most nearly

   (A) 280 in-lbf (torsion)
   (B) 280 ft-lbf (torsion)
   (C) 280 in-lbf (torsion), 500 ft-lbf (flexure)
   (D) 500 ft-lbf (flexure)

Hint: Determine the shear center of the channel.

MATERIALS

PROBLEM 41

A bridge is designed with composite steel girders spanning 30 ft. The girders are W16 × 100 Grade 50 steel shapes with a flange width of 10.4 in and are spaced 6 ft on center.

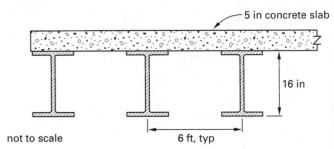

What is the effective width of slab for an interior girder?

   (A) 10 in
   (B) 60 in
   (C) 72 in
   (D) 90 in

Hint: Use AASHTO *Standard Specifications for Highway Bridges*.

PROBLEM 42

The W18 × 40 composite steel beam shown supporting a 4 in concrete slab spans 30 ft. The modular ratio, n, is 8.

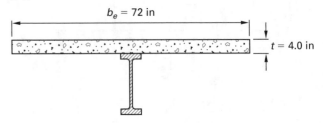

If the concrete slab is transformed to steel, the distance to the centroid of the transformed section measured from the bottom of the beam will be most nearly

   (A) 14 in
   (B) 16 in
   (C) 17 in
   (D) 19 in

Hint: The modular ratio, n, is the ratio of E_s/E_c.

PROBLEM 43

A W18 × 40 composite beam with a 4 in concrete slab supports a live load moment of 140 ft-kips and a dead load moment of 80 ft-kips. The dead load includes the weight of the slab and beam. The construction is unshored.

section modulus of the beam	68.4 in³
transformed section modulus, measured to the bottom of the section	103 in³
transformed section modulus, measured to the top of the concrete	350 in³
transformed section modulus, measured to the steel/concrete interface	1680 in³

The bending stress in the bottom fibers of the steel beam due to dead load is most nearly

- (A) 9.3 kips/in²
- (B) 14 kips/in²
- (C) 39 kips/in²
- (D) 1200 lbf/in²

Hint: In unshored construction, the beam carries the full dead load.

PROBLEM 44

The maximum bending stress in a composite Grade 50 steel beam with shear connectors, constructed without shoring, is limited to

- (A) 32 kips/in²
- (B) 33 kips/in²
- (C) 38 kips/in²
- (D) 45 kips/in²

Hint: Use Ch. I of the AISC ASD manual.

PROBLEM 45

A simply supported composite steel beam has the following properties.

area of concrete	288 in²
area of steel	11.8 in²
compressive strength of concrete	3000 lbf/in²
yield stress of steel	36 kips/in²

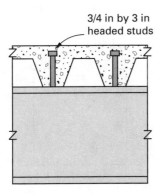

3/4 in by 3 in headed studs

The composite deck has a nominal rib height of 2 in, an average rib width of 3 in, and one stud per rib. The total number of studs needed on the beam is

- (A) 19
- (B) 29
- (C) 37
- (D) 58

Hint: Use Ch. I of the AISC ASD manual.

PROBLEM 46

A structural masonry wall is designed to span 10 ft vertically from a structure's foundation to the roof. The wall is not reinforced but is grouted solid and carries the load from the roof. The compressive strength of masonry is not specified. Lateral support is provided by intersecting walls spaced 15 ft on center. What is the minimum thickness of the wall?

- (A) 6 in
- (B) 8 in
- (C) 9 in
- (D) no limit

Hint: A structural masonry wall that supports an axial load (i.e., roof load) is a bearing wall.

PROBLEM 47

A prestressed concrete beam has a specified compressive strength of concrete of 6000 lbf/in² and the following properties.

compressive strength of concrete at time of initial prestress	4000 lbf/in²
specified yield strength of nonprestressed reinforcement	60,000 lbf/in²
initial prestress force	150 kips

The extreme fiber stress in tension in the concrete immediately after prestress transfer is limited to

- (A) 150 lbf/in²
- (B) 190 lbf/in²
- (C) 230 lbf/in²
- (D) 380 lbf/in²

Hint: Refer to ACI 318 Ch. 18.

PROBLEM 48

1800 lbf

The maximum factored shear in the 6 in by 9 in concrete beam shown is ~~2000~~ lbf. The compressive strength of concrete is 3000 lbf/in² and the yield stress of the steel reinforcement is 60,000 lbf/in².

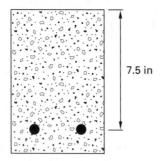

7.5 in

What is the required shear reinforcement?

(A) no. 3 U-stirrups at 3.75 in spacing
(B) no. 3 U-stirrups at 4.5 in spacing
(C) no. 3 U-stirrups at 24 in spacing
(D) none

Hint: Refer to ACI 318 Sec. 11.5.

PROBLEM 49

A 20 in by 30 in reinforced concrete beam has five continuous 20 ft spans. The point of inflection is 3 ft from the face of the support. According to ACI 318, the minimum clear cover on the reinforcing bars is 1.5 in and the minimum clear spacing of the reinforcing bars is 2.5 in.

concrete compressive strength	6000 lbf/in^2
yield stress of reinforcement	60,000 lbf/in^2

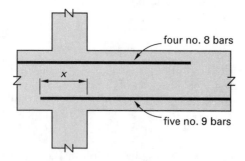

four no. 8 bars

x

five no. 9 bars

The minimum length of the bottom bars beyond the face of the support, x, is most nearly

(A) 0 in
(B) 6 in
(C) 30 in
(D) 40 in

Hint: Refer to ACI 318 Ch. 12.

PROBLEM 50

A glued laminated (glulam) beam with the designation 20F-V3 is used to span a 20 ft opening. Which of the following statements is true?

(A) The beam is mechanically graded.
(B) The depth of the beam is 20 in.
(C) The shear capacity of the beam is 300 lbf/in^2.
(D) The flexural tensile capacity of the beam is 2000 lbf/in^2.

Hint: The answer does not depend upon the type of wood.

PROBLEM 51

Southern pine glued laminated timbers (glulams) are used for rafters in a church. The rafters are uniformly loaded and have full lateral support along the compression edges. The tabulated design bending value for the tension zone stressed in tension, F_{bxx}, is 2000 lbf/in^2. The glulams have a width of $8^1/_2$ in and a depth of $27^1/_2$ in. The distance between points of zero moment for the timbers is 15 ft. The loads on each rafter are

dead load moment	60,000 ft-lbf
live load moment	90,000 ft-lbf
wind load moment	100,000 ft-lbf [downward pressure]

The local code allows a one-third stress increase for wind loading. Using the NDS, what is the critical ratio of F_b/F_b'?

(A) 0.84
(B) 0.88
(C) 0.92
(D) 1.1

Hint: Two load cases must be considered—dead load plus live load, and dead load plus live load plus wind load.

MEMBER DESIGN
PROBLEM 52

A 7.5 in wide by 16 in deep reinforced concrete masonry lintel is used to carry uniform dead and live loads of 450 lbf/ft and 550 lbf/ft, respectively. The self-weight of the lintel is 120 lbf/ft. The load due to the weight of the wall above the lintel is as shown for the lintel span length, L. The specified compressive strength of the masonry is 3000 lbf/in^2. Design the lintel using allowable stress design.

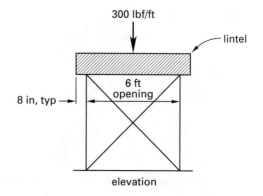

300 lbf/ft

lintel

8 in, typ

6 ft opening

elevation

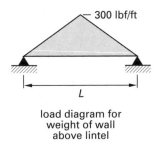

load diagram for
weight of wall
above lintel

If two no. 4, Grade 60 reinforcing bars are used side-by-side at the bottom of the beam, what is the tensile stress in the steel?

(A) 13,900 lbf/in^2
(B) 14,900 lbf/in^2
(C) 15,600 lbf/in^2
(D) 17,200 lbf/in^2

Hint: Calculate the lintel span length first.

PROBLEM 53

A 35 in by 40 in reinforced concrete beam has longitudinal and torsional reinforcement with a yield stress of 40,000 lbf/in^2. The compressive strength of the concrete is 4000 lbf/in^2. The nominal shear strength of the concrete is 168.2 kips. The area enclosed by the outside perimeter of concrete is 1400 in^2. The outside perimeter of concrete cross section is 150 in.

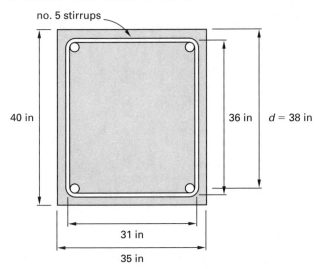

The torsion and shear diagrams are as shown.

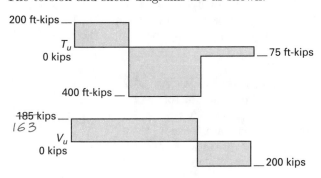

What is the maximum spacing of the no. 5 stirrups shown at the point of maximum torsion?

(A) 2.2 in   2.0 in
(B) 3.2 in   2.7 "
(C) 3.4 in   3.1 "
(D) 4.2 in   3.7 "

Hint: Refer to ACI 318 Sec. 11.6 for beams with torsion.

PROBLEM 54

A 35 in by 40 in, 4000 lbf/in^2 reinforced concrete beam has yield strengths of 60,000 lbf/in^2 for the longitudinal torsional reinforcement and 40,000 lbf/in^2 for the closed transverse torsional reinforcement. The spacing of the torsional reinforcement is 0.0744 in^2/in. The factored torsional load on the beam is 400 ft-kips. The positive factored moment on the beam is 500 ft-kips. The perimeter of the area enclosed by the centerline of the transverse reinforcement is 134 in.

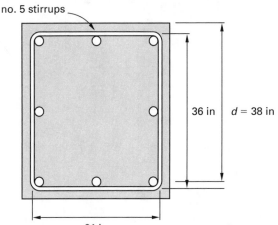

the reduction factor for torsional reinforcement in the flexural compression zone is not applied and

If a minimum number of bars is used, the total area of longitudinal reinforcement in the bottom of the beam is most nearly

(A) three no. 11 bars
(B) four no. 11 bars
(C) five no. 11 bars
(D) four no. 14 bars

Hint: Refer to ACI 318 Sec. 11.6 for design of beams with torsion.

PROBLEM 55

A 6 in normal-weight concrete flat slab supports a factored load of 320 lbf/ft^2. The beam shown is a continuous two-span beam between columns.

specific weight of slab	150 lbf/ft^3
α_1	1.5
positive moment in the column strip	90 ft-kips
negative moment in the column strip	−220 ft-kips

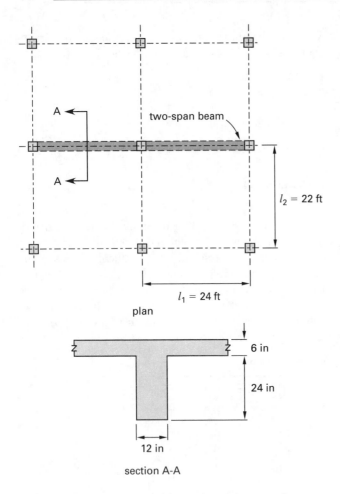

plan

section A-A

What is the maximum midspan beam moment?

- (A) 77 ft-kips
- (B) 89 ft-kips
- (C) 94 ft-kips
- (D) 290 ft-kips

Hint: Refer to ACI 318 Ch. 13.

PROBLEM 56

A $4 \times 4 \times \frac{3}{8}$ tube beam is welded to the flange of a W 10×33 steel column as shown. The reaction at the column is 12,000 lbf. A36 steel is used for both the column and the beam. The welds are fillet welds made using the shielded metal arc process (SMAW) with E70XX electrodes.

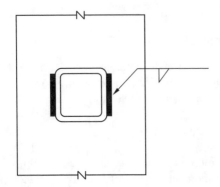

What is the required size of the weld?

- (A) $\frac{1}{8}$ inch
- (B) $\frac{3}{16}$ inch
- (C) $\frac{1}{4}$ inch
- (D) The shear exceeds the capacity of a fillet weld.

Hint: Use Part 5, Sec. J2 of the AISC ASD manual.

PROBLEM 57

A 60 kip load is suspended from the bottom flange of a W30 × 90 beam. An equal-leg double-angle hanger (A36 steel) is used to support the load. The hanger is attached to the beam with four $\frac{7}{8}$ in diameter A325-N bolts as shown.

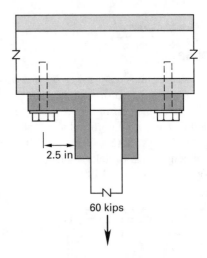

Size the double-angle hanger.
- (A) L5 × 5 × $\frac{1}{2}$
- (B) L5 × 5 × $\frac{7}{8}$
- (C) L6 × 6 × 1
- (D) L8 × 8 × $1\frac{1}{8}$

Hint: Refer to Part 4 of the AISC ASD manual.

PROBLEM 58

The tension member shown is constructed with two 2×6 pieces of red oak. Four rows of three $\frac{1}{2}$ in diameter bolts are used to connect the pieces. The spacing between the rows is 0.75 in. The spacing between the closest fasteners, measured parallel to the rows, is 3.25 in.

modulus of elasticity	1.2×10^6 lbf/in^2
load duration factor	1.0
wet service factor	1.0
temperature factor	1.0
geometry factor	1.0
end grain factor	1.0

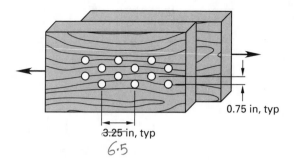

0.75 in, typ

3.25 in, typ

6.5

What is the capacity of the connection?

(A) 610 lbf
(B) 5800 lbf
(C) 7300 lbf
(D) 7700 lbf

Hint: Refer to NDS Ch. 10.

PROBLEM 59

A W14 × 22 steel beam (A36) is bolted to a column flange with L3$^1/_2$ × 3$^1/_2$ × $^5/_{16}$ double angles as shown. A single row of $^3/_4$ in diameter A325-N bolts is used.

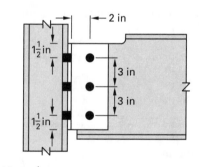

not to scale

Using the AISC ASD manual, the maximum beam reaction is most nearly

(A) 12 kips
(B) 33 kips
(C) 45 kips
(D) 56 kips

Hint: Refer to Part 4 of the AISC ASD manual for connection design.

PROBLEM 60

An L4 × 8 angle (LLV) is welded to a column using the shielded metal arc process and E70XX electrodes.

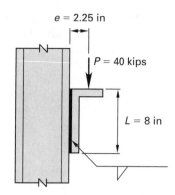

$e = 2.25$ in

$P = 40$ kips

$L = 8$ in

What is the required fillet weld size if the angle is welded on both sides of the vertical leg only?

(A) $^3/_{16}$ in
(B) $^1/_4$ in
(C) $^3/_8$ in
(D) $^1/_2$ in

Hint: The weld is subject to both shear and bending.

PROBLEM 61

A W14 × 109 steel column carries an axial load of 320 kips and a moment of 200 ft-kips about the y-axis. The centerline of the anchor bolts is 1.5 in outside the column flanges as shown.

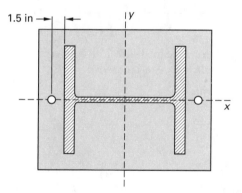

1.5 in

If A307 anchor bolts are used, what size bolt is needed?

(A) $^5/_8$ in diameter
(B) $^3/_4$ in diameter
(C) 1 in diameter
(D) 1$^1/_8$ in diameter

Hint: Columns with relatively large moments may be subject to uplift on the base plate.

PROBLEM 62

Plywood sheathing is attached to 2 × 12 roof rafters with 6d box nails. The nails penetrate into the rafters 1 in. The rafters are spaced 16 in on center. All members are southern pine. Sustained temperatures do not exceed 100°F. If the uplift on the roof is 36 lbf/ft², the maximum nail spacing is most nearly

    (A) 7.8 in
    (B) 11 in
    (C) 12 in
    (D) 14 in

Hint: Refer to NDS Ch. 11.

PROBLEM 63

A continuous L4 × 6 × ³/₈ angle anchored to a 12 in concrete masonry wall is used as a ledger for floor joists as shown. A36 anchor bolts are placed 16 in on center in grouted cells. The joists are spaced 6 ft on center. The specified compressive strength of masonry is 1500 lbf/in².

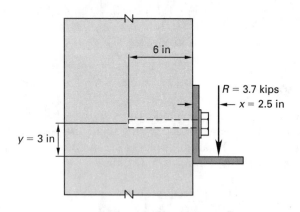

If the load from the joists is 3.7 kips, what is the smallest diameter bolt that can be used?

    (A) ³/₈ in
    (B) ¹/₂ in
    (C) ⁵/₈ in
    (D) ³/₄ in

Hint: The bolt is subject to both shear and tension.

PROBLEM 64

A built-up column made from three southern pine sawn 2 × 6 members nailed together meets the requirements of NDS Sec. 15.3.3. The ends of the column are free to rotate but are not free to translate.

distance between points of lateral support of compression member in plane 1	9.0 ft
distance between points of lateral support of compression member in plane 2	4.5 ft
allowable modulus of elasticity	1.7×10^6 lbf/in²
F_c^*	1750 lbf/in²
K_{cE}	0.3

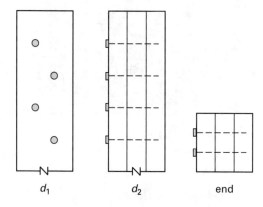

What is the critical column stability factor for this column?

    (A) 0.44
    (B) 0.52
    (C) 0.57
    (D) 0.59

Hint: NDS Sec. 15.3 covers the design of built-up columns.

PROBLEM 65

A 20 in by 24 in short concrete column carries an axial factored load of 750 kips and a factored moment of 600 ft-kips. Bars are located on the short faces of the columns and γ is 0.75. The concrete has a compressive strength of 4000 lbf/in² and the reinforcement has a yield stress of 60,000 lbf/in². The required area of steel is most nearly

    (A) 12 in²
    (B) 14 in²
    (C) 19 in²
    (D) 36 in²

Hint: Use an interaction diagram to determine the required area of steel.

PROBLEM 66

A reinforced concrete building has a 6 in flat-plate slab on each floor and 12 in diameter columns. The floor-to-floor distance is 13 ft. A column has a factored axial load of 300 kips and equal end moments of 100 ft-kips and is subjected to double curvature. The modulus of elasticity of concrete is 3.6×10^6 lbf/in². If the column does not have any transverse loads and is not subject to sway, the design moment is most nearly

    (A) 100 ft-kips
    (B) 4000 ft-kips
    (C) 9000 ft-kips
    (D) 20,000 ft-kips

Hint: Use magnified moments to determine the design moment.

PROBLEM 67

A three-story reinforced concrete building is supported on columns placed on a grid and spaced 20 ft in each direction. The columns are 18 in by 18 in. Beams with an overall depth of 20 in and a web width of 10 in span between the columns in each direction. The slabs are all 6 in thick and are two-way spans. The moment of inertia of the beams for the computation of the relative stiffness parameter, Ψ, at the top of the first-floor corner column is most nearly

(A) 3400 in^4
(B) 3700 in^4
(C) 9800 in^4
(D) 12,000 in^4

Hint: The relative stiffness parameter is used in the determination of the effective length factor, k.

PROBLEM 68

A three-story reinforced concrete building is supported on columns placed on a grid and spaced 20 ft apart in each direction. The story height, measured from the top of one slab to the top of the slab above, is 13 ft. At the first level, the distance from the tops of the footings to the top of the first-story slab is 18 ft. The columns are 18 in by 18 in. Beams span between columns in each direction. The slabs are all 6 in thick two-way spans. 6000 lbf/in^2 concrete with a modulus of elasticity of 4.4×10^6 lbf/in^2 is used throughout the structure.

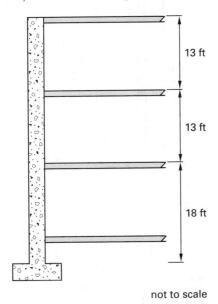

13 ft

13 ft

18 ft

not to scale

elevation

If the moment of inertia of the spandrel beams is 10,000 in^4, what is the relative stiffness parameter, Ψ, at the top of the first-floor corner column?

(A) 2.3
(B) 4.6
(C) 5.4
(D) 6.6

Hint: Refer to ACI 318 Ch. 10.

PROBLEM 69

A six-story reinforced concrete building is supported on columns placed on a grid and spaced 20 ft apart in each direction. The story height, measured from the top of one slab to the top of the slab above, is 13 ft. At the first level, the distance from the tops of the footings to the top of the first story slab is 18 ft. The columns are 18 in by 18 in (the moment of inertia is 8800 in^4). Beams span between columns in each direction. Their moment of inertia is 10,000 in^4. The slabs are all 6 in thick two-way spans. 6000 lbf/in^2 concrete with a modulus of elasticity of 4.4×10^6 lbf/in^2 is used throughout the structure. What is the effective length in the East/West direction for an interior column in an unbraced frame at the second story?

(A) 8.5 ft
(B) 18 ft
(C) ~~22 ft~~ 21 ft
(D) 23 ft

Hint: Determine the relative stiffness parameter for the column.

PROBLEM 70

A 20 in by 20 in brick masonry column with a specified compressive strength of masonry of 3500 lbf/in^2 is reinforced with four no. 6 bars. It carries a compressive load with 5.0 in eccentricity in the x direction and 6.2 in eccentricity in the y direction. Compression controls the design.

ratio of effective height to radius of gyration (h/r)	72
steel spacing ratio, g	0.4

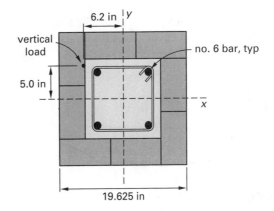

The maximum biaxial load is most nearly

(A) 54,000 lbf
(B) 96,000 lbf
(C) 100,000 lbf
(D) 250,000 lbf

Hint: The maximum biaxial load is not a linear sum of the individual capacities.

PROBLEM 71

A 4.5 ft by 20 ft rectangular concrete footing supports two 12 in concrete columns.

Column 1:
116
    dead load   ~~100~~ kips
    live load   ~~60~~ kips
64
Column 2:
70
    dead load   ~~60~~ kips
    live load   ~~30~~ kips
32

yield stress of reinforcement	$60{,}000 \text{ lbf/in}^2$
compressive strength of concrete	3000 lbf/in^2
depth to reinforcement	15 in

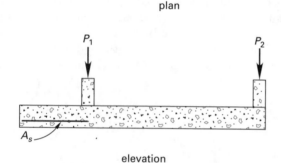

plan

elevation

How many no. 6 bars are required in the longitudinal direction between column 1 and the edge of the footing if the critical design moment is located at the outer face of column 1? $\longrightarrow$ a soil pressure is assumed to be uniform

    (A) two
    (B) six
    (C) seven
    (D) eight

Hint: Concrete footing design is based on factored loads.

PROBLEM 72

A rectangular concrete footing with a compressive strength of 3 kips/in² supports a centrally located column carrying a factored load of 60 kips. The column is 12 in square and the footing is 5 ft by 10 ft. The depth to the reinforcement is 12 in. The yield stress of the reinforcement is 60 kips/in². What is the area of steel required in the short direction of the footing under the column?

    (A) five no. 4 bars
    (B) seven no. 4 bars
    (C) nine no. 4 bars
    (D) nine no. 5 bars

Hint: A footing having a centrally located column with only an axial load has a uniform pressure distribution.

PROBLEM 73

The 6 in nominal (5⅝ in actual), exterior, nonloadbearing reinforced concrete masonry wall shown is subjected to 20 lbf/ft² wind pressure. The wind produces a reaction at the roof of 163 lbf/ft and a reaction at the foundation of 117 lbf/ft. The maximum moment produced by the wind is 341 ft-lbf/ft.

compressive strength of masonry	1500 lbf/in^2
modular ratio	21.5

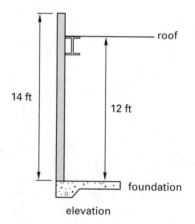

elevation

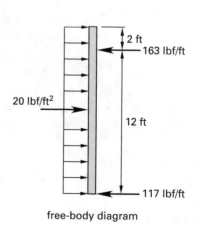

free-body diagram

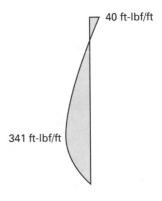

40 ft-lbf/ft

341 ft-lbf/ft

moment diagram

Using a bar spacing of 32 in and Grade 60 reinforcing bars, what is the allowable moment based on the flexural compressive stress in the masonry?

    (A) 5700 in-lbf/ft
    (B) 6100 in-lbf/ft
    (C) 7600 in-lbf/ft
    (D) 11,000 in-lbf/ft

Hint: Start by determining the required area of reinforcement.

PROBLEM 74

A deck is built on the back of a house in a humid climate using the design requirements contained in the NDS. A 2×8 beam is screwed to the face of a 4×4 post as shown. The wood is southern pine. The screws shall have a penetration greater than or equal to ~~seven~~ TEN times the shank diameter. The dead-load plus live-load end reaction of the beam is 605 lbf.

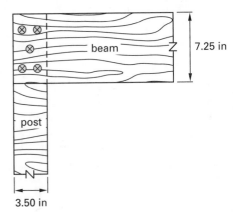

beam

7.25 in

post

3.50 in

If five 12-gage wood screws are used, what is the allowable load on the connection?

    (A) 450 lbf
    (B) 500 lbf
    (C) 560 lbf
    (D) 800 lbf

Hint: Refer to NDS Ch. 11 for wood screw design values.

PROBLEM 75

A beam-column made from Grade 50 steel supports an axial load of 400 kips at an eccentricity of 12 in about the strong axis. The effective length, KL, is 25 ft. Using the AISC ASD manual, select a W shape to carry the load.

    (A) $W14 \times 193$
    (B) $W14 \times 233$
    (C) $W14 \times 257$
    (D) $W14 \times 342$

Hint: Use the equivalent axial load procedure found in Part 3 of the AISC ASD manual.

PROBLEM 76

Wood formwork consisting of 3×4 joists and stringers is used to support a 4 in normal-weight concrete slab. Given that the maximum spacing of the stringers limited by bending in the joists is $(4466 \text{ lbf}/w)^{1/2}$, the maximum spacing of the stringers as limited by bending in the stringers is $(523.4 \text{ lbf/ft})/w$ and the maximum spacing of the stringers as limited by the load to the post is $(512.8 \text{ lbf/ft})/w$. The maximum spacing of the stringers is most nearly

    (A) 5.1 ft
    (B) 5.9 ft
    (C) 6.7 ft
    (D) 9.5 ft

Hint: Determine the load on the stringers.

PROBLEM 77

Which of the following statements are true?

I.   Mat foundations can be used in areas where the basement is below the GWT.

II. Mat foundations always require a top layer of reinforcing bars.

III. Conventional spread footings tolerate larger differential settlements than do mat foundations.

IV. Mat foundations are not suitable where settlement may be a problem.

    (A) I
    (B) I and II
    (C) III and IV
    (D) II, III, and IV

Hint: A mat foundation is a large concrete slab in contact with the soil and is commonly used to support several columns or pieces of equipment.

PROBLEM 78

The cantilevered sheet piling shown has a concentrated lateral load, H, of 10 kips spaced 2 ft on center, horizontally along the top of the piling. The soil specific weight is 110 lbf/ft^3 and the angle of internal friction of the soil is 30°.

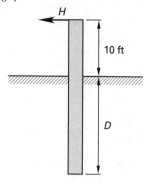

If the factor of safety for the minimum depth is 1.3, the total length of the sheet pile is most nearly

(A) 21 ft
(B) 26 ft
(C) 31 ft
(D) 38 ft

Hint: The solution for this problem is an iterative one.

PROBLEM 79

An HP12 × 63 steel pile is driven 100 ft into saturated soft clay. The cohesion, c, is 400 lbf/ft^2 and the adhesion, c_A, is 360 lbf/ft^2. What is the allowable bearing capacity of the pile if the factor of safety is 3?

(A) 1.7 kips
(B) 49 kips
(C) 69 kips
(D) 150 kips

Hint: The allowable bearing capacity is the sum of the point-bearing capacity and the skin-friction capacity divided by the factor of safety.

PROBLEM 80

A wooden pier is supported on kiln-dried round timber piles that extend 20 ft below the water surface into clay soil. A group of three piles carries a 300 kip dead load and a 400 kip live load. The piles are made of red oak and have an area of 230 in^2. The column stability factor is 0.62. What is the total adjustment factor for the critical compression load case?

(A) 0.551
(B) 0.619
(C) 0.620
(D) 0.688

Hint: Refer to the NDS.

PROBLEM 81

A normal-weight (0.150 kip/ft^3) reinforced concrete retaining wall is designed to support the loads shown. An 8 ft wide concrete walkway extends along the entire

length of the retaining wall. The soil behind the retaining wall is soil 1. The soil in front of the retaining wall is soil 2.

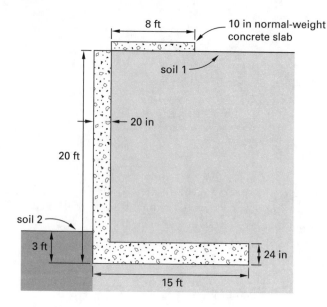

characteristic	soil 1	soil 2
unit weight (kips/ft^3)	0.110	0.100
angle of internal friction (degrees)	28	15
cohesion (kips/ft^2)	0	0.300
active earth pressure coefficient	0.361	0.589
passive earth pressure coefficient	2.77	1.70

The factor of safety against overturning is

(A) 4.15
(B) 4.24
(C) 4.64
(D) 4.86

Hint: The factor of safety against overturning is the ratio of the resisting moment to the overturning moment.

PROBLEM 82

A normal-weight (0.150 kip/ft^3) reinforced concrete retaining wall is designed to support the loads shown. An 8 ft wide concrete walkway extends along the entire length of the retaining wall. The soil is sandy without silt. The soil behind the retaining wall is soil 1. The soil in front of the retaining wall is soil 2.

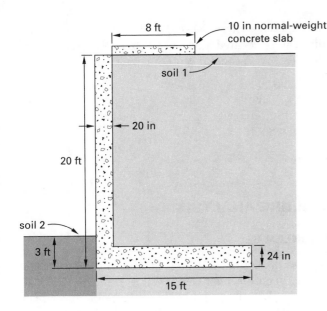

characteristic	soil 1	soil 2
unit weight (kips/ft^3)	0.110	0.100
angle of internal friction (degrees)	28	15
cohesion (kips/ft^2)	0	0.300
active earth pressure coefficient	0.361	0.589
passive earth pressure coefficient	2.77	1.70

The factor of safety against sliding is

(A) 2.20
(B) 2.26
(C) 2.45
(D) 2.61

Hint: Passive restraint from the soil is only considered if it will always be there. In most cases it is neglected.

PROBLEM 83

The reinforced concrete shear wall shown is made from concrete with a compressive strength of 4000 lbf/in^2.

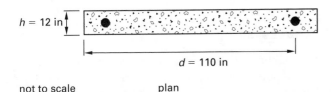

$h = 12$ in

$d = 110$ in

not to scale   plan

The nominal shear strength provided by the shear reinforcement is 700 kips. The factored axial gravity load on the wall is 200 lbf/ft. Seismic loads can be ignored. The nominal shear strength of the wall is most nearly

(A) 170 kips
(B) 710 kips
(C) 830 kips
(D) 870 kips

Hint: Refer to ACI 318 Sec. 11.10.

PROBLEM 84

A 30 ft high, 10 in thick reinforced concrete shear wall has the following properties.

compressive strength of concrete	6000 lbf/in^2
yield stress of reinforcement	60,000 lbf/in^2
depth to reinforcement	110 in
horizontal length of wall	120 in

The factored shear force of 110 kips is due to wind loads only. If a single mat of reinforcement is used, the horizontal shear reinforcement required is most nearly

(A) no. 3 at 12 in
(B) no. 3 at 18 in
(C) no. 5 at 12 in
(D) no. 5 at 15 in

Hint: Refer to ACI 318 Sec. 11.10.

PROBLEM 85

A single-story concrete bearing wall supported at the roof and foundation has the following properties.

wall thickness	12 in
concrete compressive strength	4000 lbf/in^2
effective length factor	1.0
length of wall	16 ft
uniform factored axial load	5 kips/ft

If the wall is empirically designed, its maximum height is most nearly

(A) 25.0 ft
(B) 27.3 ft
(C) 31.6 ft
(D) 32.0 ft

Hint: Refer to ACI 318 Ch. 14.

PROBLEM 86

A 10 ft high, 30 ft long reinforced concrete shear wall has a compressive strength of 4000 lbf/in^2 and a steel reinforcement yield stress of 60,000 lbf/in^2. The ratio of horizontal shear reinforcement area to gross concrete area, ρ_h, is 0.0040. The ratio of vertical shear reinforcement area to gross concrete area should be

(A) 0.00212
(B) 0.00250
(C) 0.00400
(D) 0.00413

Hint: Refer to ACI 318 Sec. 11.10.9.

PROFESSIONAL PUBLICATIONS, INC.

PROBLEM 87

An unreinforced concrete masonry shear wall is subject to a combined axial load from the floor and roof and in-plane bending from lateral wind forces. The wall is 20 ft high and 17.7 ft long. It is built with 12 in hollow concrete masonry units with a weight of 51.0 lbf/ft². The specified compressive strength of masonry is 2500 lbf/in². The wind load, W, of 3200 lbf acts at a point 18 ft from the base of the wall.

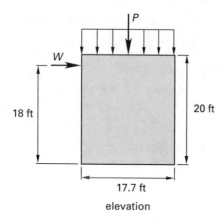

elevation

Ignoring the allowable stress increase for wind loads, the minimum and maximum allowable applied axial loads are most nearly

(A) 0.0 lbf and 380,000 lbf
(B) 1500 lbf and 370,000 lbf
(C) 1500 lbf and 1,500,000 lbf
(D) 20,000 lbf and 380,000 lbf

Hint: Refer to Ch. 2 of the MSJC code.

PROBLEM 88

A two-way flat-slab system is supported by 12 in concrete columns as shown.

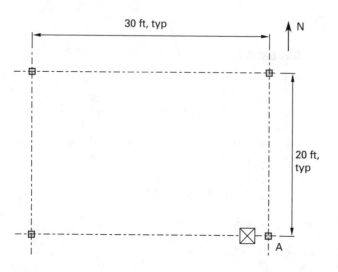

If no special analysis is used, the maximum size opening that can be located adjacent to column A, centered on the East/West column centerline, is most nearly

(A) 0.0 ft²
(B) 0.39 ft²
(C) 1.6 ft²
(D) 2.3 ft²

Hint: Refer to ACI 318 Ch. 13, on two-way slabs.

FAILURE ANALYSIS

PROBLEM 89

A truss made from A36 steel is used to carry vehicular traffic. The diagonal compression members are made from two L8 × 6 × ½ angles with a ½ in gusset plate between them as shown. The short legs of the angles are back-to-back. The controlling slenderness ratio, Kl/r, is 110. Use the AISC ASD manual.

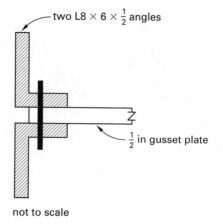

not to scale

If the axial load is concentric, what is the allowable compressive stress?

(A) 6.85 kips/in²
(B) 11.2 lbf/in²
(C) 11.2 kips/in²
(D) 11.7 lbf/in²

Hint: The long legs of the angle are unstiffened elements.

PROBLEM 90

A steel channel strut with bolted bearing end connections is subjected to a dead load tensile force of 20 kips. The live load varies from a 10 kip compressive load to a 50 kip tensile force, and it is estimated that the live load may be applied 200 times per day for the life of the structure. The structure is expected to last at least 25 years. If A36 steel is used, what is the lightest section that can carry the load?

(A) C7 × 9.8
(B) C6 × 13
(C) C8 × 11.5
(D) C8 × 18.75

Hint: Reversal of the live load must be considered in the design.

DESIGN CRITERIA

PROBLEM 91

An elementary school is constructed with hollow brick bearing walls with a specified compressive strength of masonry of 4500 lbf/in². The brick meets the requirements of ASTM specification C652, and Type S mortar is used. What is the modulus of elasticity of the masonry?

(A) 1.3×10^6 lbf/in²
(B) 2.6×10^6 lbf/in²
(C) 3.2×10^6 lbf/in²
(D) 4.0×10^6 lbf/in²

Hint: ASTM specification C652 is for clay brick.

PROBLEM 92

A circular 50 ft high, 20 ft diameter water tank provides drinking water to a seaside resort town on the Pacific Ocean. The basic wind speed for the town is 85 mph. The terrain is flat. The tank is made from concrete, is designed as a rigid structure, and is moderately smooth. Using the IBC, the design wind force on the leeward side of the tank is

(A) 24.7 lbf/ft²
(B) 3030 lbf
(C) 3560 lbf
(D) 3700 lbf

Hint: The simplified method cannot be used in this case.

PROBLEM 93

A two-story wood-framed apartment building is classified as Seismic Design Category C, according to the IBC. Which of the following statements about this building is FALSE?

(A) The total design lateral seismic force increases as the building weight increases.
(B) The short-period response accelerations for this site must be between $0.33g$ and $0.50g$.
(C) There is no limit on story drift.
(D) When soil properties are not known in sufficient detail to determine the site class, Site Class D should be used unless determined otherwise by the building official.

Hint: Refer to IBC Sec. 1617.

PROBLEM 94

The plain concrete foundation wall shown supports a 12 in concrete masonry unit (CMU) wall. The foundation wall bears on soil composed of poorly graded clean sands & is laterally supported @ the top & bottom.

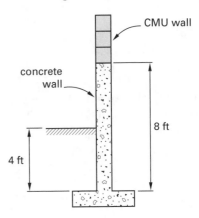

Using the prescriptive criteria found in the IBC, what is the minimum thickness of the foundation wall?

(A) 7.5 in
(B) 8 in
(C) 10 in
(D) 12 in

Hint: Use IBC Sec. 1805 to determine the minimum thickness.

PROBLEM 95

A single-story steel-framed building has columns spaced 15 ft on center in the North/South direction and 20 ft on center in the East/West direction. The columns support a uniform dead load of 40 lbf/ft² and a uniform live load of 40 lbf/ft². The building has a roof with a 1:2 pitch. Using the IBC, the minimum live load on an interior column is most nearly

(A) 4.9 kips
(B) 9.7 kips
(C) 12 kips
(D) 24 kips

Hint: IBC Sec. 1607 covers live loads.

PROBLEM 96

The roof of a multistory building is steel composed of steel joists spaced 5 ft on center and spanning 30 ft between steel girders. The girders span 50 ft between columns. The loads on the flat roof are

roof dead load   32 lbf/ft²
roof live load   20 lbf/ft²

Using the IBC, what is most nearly the live load on an interior roof girder?

(A) 12 kips
(B) 18 kips
(C) 26 kips
(D) 30 kips

Hint: Refer to IBC Ch. 16, Structural Loads.

PROBLEM 97

A church is built with a wood roof that is exposed on the interior. The roof beam is 60 ft in length. According to the IBC, what is the total deflection limit for the roof beam?

(A) 0.5 in
(B) 3 in
(C) 4 in
(D) 6 in

Hint: Refer to IBC Sec. 1604.3.

PROBLEM 98

A reinforced brick masonry residence located in Los Angeles, California, is designed using strength design provisions. According to the MSJC code, which of the following requirements must be met during design and construction of this building?

I.   Verify placement of reinforcement prior to grouting.
II. Verify placement of grout continuously during construction.
III. Observe preparation of mortar specimens.
IV. Verify the compressive strength of masonry prior to construction and every 5000 ft^2 during construction.

(A) I and II
(B) I and III
(C) I, II, and III
(D) none of the above

Hint: These requirements are part of a quality assurance program.

PROBLEM 99

Which of the following statements is/are FALSE regarding the NDS?

I.   The tabulated allowable bending design values include the effects of the beam stability factor.
II. The tabulated allowable design values for structural glued laminated timber (glulam) include adjustments for size.

III. The tabulated allowable design values for round timber piles include adjustments to compensate for the strength reduction associated with untreated piles.
IV. The group action factor applies to all types of wood connections (i.e., nails, bolts, spikes).

(A) I
(B) II
(C) I and IV
(D) I, III, and IV

Hint: Use the applicability tables in the NDS to determine which factors apply.

PROBLEM 100

A loadbearing ~~brick~~ masonry ~~fire station~~ _warehouse_ is located in central Florida. The building has a flat roof supported by steel joists and metal decking. The roof height is 30 ft plus a 2 ft high parapet. The roof diaphragm is part of the main wind-resisting force system (MWRFS). The building is 40 ft by 90 ft in plan and is classified as ~~partially~~ enclosed ~~because of the large equipment doors~~. The structure is designed using allowable stress design. The roof live load is 20 lbf/ft^2. The roof dead loads include the following components.

joists	5 lbf/ft^2
roof deck	3 lbf/ft^2
rigid insulation	3 lbf/ft^2
felt and gravel	5 lbf/ft^2

The basic wind speed is 100 mph. Using the IBC provisions, the critical design load on the roof deck portion of the MWRFS is most nearly

(A) ~~15~~ lbf/ft^2 (uplift) 13
(B) ~~22~~ lbf/ft^2 (uplift) 19
(C) 31 lbf/ft^2
(D) 45 lbf/ft^2

Hint: Be sure to consider all load combinations.

Breadth Solutions

LOADINGS

SOLUTION 1

The load on the beam, w, is the weight of the wall per foot.

$$w = \gamma h t$$
$$= \left(85 \ \frac{\text{lbf}}{\text{ft}^3}\right)(8 \ \text{ft})(6 \ \text{in})\left(\frac{1 \ \text{ft}}{12 \ \text{in}}\right)$$
$$= 340 \ \text{lbf/ft}$$

The answer is (A).

Why Other Options Are Wrong

(B) This incorrect solution results from failing to multiply by the wall width.

(C) This incorrect solution is a result of calculating the total load rather than the load per foot.

(D) This incorrect solution fails to perform the conversion from inches to feet.

SOLUTION 2

The tributary width for an interior column is found by summing the halved distances to the adjacent column(s) from the column centerline.

$$w' = \sum \tfrac{1}{2}d$$
$$w_1' = \left(\frac{1}{2}\right)(20 \ \text{ft}) + \left(\frac{1}{2}\right)(20 \ \text{ft})$$
$$= 20 \ \text{ft}$$
$$w_2' = \left(\frac{1}{2}\right)(20 \ \text{ft}) + \left(\frac{1}{2}\right)(20 \ \text{ft})$$
$$= 20 \ \text{ft}$$

The tributary area for an interior column is

$$A = w_1' w_2'$$
$$= (20 \ \text{ft})(20 \ \text{ft})$$
$$= 400 \ \text{ft}^2$$

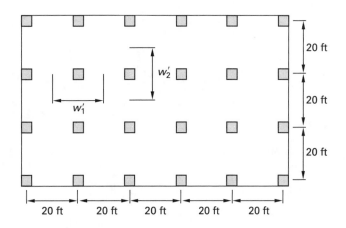

The total uniform load per unit area is

$$w_{\text{uniform}} = (D_{\text{roof}} + L_{\text{roof}}) + (D_{\text{floor}} + L_{\text{floor}})$$
$$= \left(15 \ \frac{\text{lbf}}{\text{ft}^2} + 20 \ \frac{\text{lbf}}{\text{ft}^2}\right) + \left(15 \ \frac{\text{lbf}}{\text{ft}^2} + 60 \ \frac{\text{lbf}}{\text{ft}^2}\right)$$
$$= 110 \ \text{lbf/ft}^2$$

The total column load is

$$P = w_{\text{uniform}} A$$
$$= \left(110 \ \frac{\text{lbf}}{\text{ft}^2}\right)(400 \ \text{ft}^2)$$
$$= 44{,}000 \ \text{lbf} \quad (44 \ \text{kips})$$

The answer is (C).

Why Other Options Are Wrong

(A) This incorrect solution uses tributary width instead of tributary area when calculating the total load on the column.

(B) This incorrect solution does not include the second-floor loads when calculating the total load on the column.

(D) This incorrect solution uses factored loads, which are only applicable to concrete design or strength design. This solution was calculated for steel using the load factors in the LFRD; that is, $U = 1.2D + 1.6L$.

ANALYSIS

SOLUTION 3

From the sum of the moments about E, the vertical reaction at support A is

$$R_{A,v} = \frac{(300 \text{ lbf})(2h) + (200 \text{ lbf})(3h)}{4h}$$
$$= 300 \text{ lbf}$$

Draw the free-body diagram of joint A.

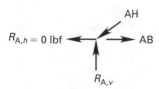

By summing the forces in the horizontal direction, determine that the horizontal reaction at A is zero.

The sum of the forces in the vertical direction is

$$AH_v - R_{A,v} = 0 \text{ lbf}$$
$$AH_v = 300 \text{ lbf (compression)}$$

From the geometry of the truss, the horizontal component of AH must be twice the vertical component.

$$AH_h = (2)(300 \text{ lbf})$$
$$= 600 \text{ lbf (compression)}$$

The resultant force in member AH is

$$AH = \sqrt{AH_v^2 + AH_h^2}$$
$$= \sqrt{(300 \text{ lbf})^2 + (600 \text{ lbf})^2}$$
$$= 670.8 \text{ lbf} \quad \left(670 \text{ lbf (compression)}\right)$$

The answer is (C).

Why Other Options Are Wrong

(A) This incorrect solution finds only the vertical component of the force in AH.

(B) This incorrect solution finds the horizontal component of the force in AH.

(D) This incorrect solution identifies the resultant force in AH as a tensile force.

SOLUTION 4

To determine what loads need to be considered on a lintel, first determine if arching action will occur. A masonry wall will exhibit arching action over an opening if sufficient masonry extends on both sides of the opening to resist the thrusting action of the arch. An extent of masonry on each side greater than the distance of the opening itself is usually adequate to resist the thrust. When arching action occurs, only the loads within a 45° triangle over the opening will be carried by the lintel.

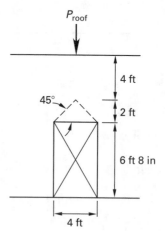

In this case, the concentrated load from the roof truss occurs beyond the apex of the 45° triangle and therefore is not included in the loads on the lintel.

Calculate the loads on the lintel.

The span length of the lintel is the center-of-bearing to center-of-bearing distance—the clear span plus one-half the bearing length on each side of the opening. The MSJC code specifies, in Sec. 2.3.3.4, a minimum bearing length of 4 in for beams. The span length is

$$L = \text{clear span} + \frac{1}{2}\sum \text{bearing length at each end}$$
$$= 4 \text{ ft} + \left(\frac{1}{2}\right)(4 \text{ in} + 4 \text{ in})\left(\frac{1 \text{ ft}}{12 \text{ in}}\right)$$
$$= 4.33 \text{ ft}$$

Determine the lintel self-weight. From the AISC ASD Table of Dimensions and Properties, the weight of the two 5 × 3½ × ½ angles is

$$w_{\text{lintel}} = (2)\left(12.0 \text{ } \frac{\text{lbf}}{\text{ft}}\right) = 24 \text{ lbf/ft}$$

(handwritten: 7/16)

The weight of the masonry above the lintel is a triangular load. At its maximum, this load is

$$w_{\text{peak}} = w_{\text{uniform}}\left(\frac{L}{2}\right)$$
$$w_{\text{uniform}} = t\gamma = (8 \text{ in})\left(\frac{1 \text{ ft}}{12 \text{ in}}\right)\left(150 \text{ } \frac{\text{lbf}}{\text{ft}^3}\right)$$
$$= 100 \text{ lbf/ft}^2$$
$$w_{\text{peak}} = \left(100 \text{ } \frac{\text{lbf}}{\text{ft}^2}\right)\left(\frac{4.33 \text{ ft}}{2}\right)$$
$$= 216.5 \text{ lbf/ft}$$

The total triangular load is

$$W = \tfrac{1}{2}Lw_{\text{peak}}$$
$$= \left(\frac{1}{2}\right)(4.33 \text{ ft})\left(216.5 \; \frac{\text{lbf}}{\text{ft}}\right)$$
$$= 469 \text{ lbf}$$

The maximum moment on the lintel is

$$M = \frac{w_{\text{lintel}}L^2}{8} + \frac{WL}{6}$$
$$= \frac{\left(24 \; \frac{\text{lbf}}{\text{ft}}\right)(4.33 \text{ ft})^2}{8} + \frac{(469 \text{ lbf})(4.33 \text{ ft})}{6}$$
$$= 395 \text{ ft-lbf} \quad (400 \text{ ft-lbf})$$

The answer is (B).

Why Other Options Are Wrong

(A) This incorrect solution uses the clear span for the span length of the lintel.

(C) This incorrect solution uses the clear span for the span length of the lintel and, although it correctly calculates the triangular load, it also mistakenly includes the load from the roof truss in the load on the lintel.

(D) This incorrect solution correctly calculates the triangular load on the lintel but mistakenly includes the load from the roof truss.

SOLUTION 5

The free-body diagram for the beam described is shown.

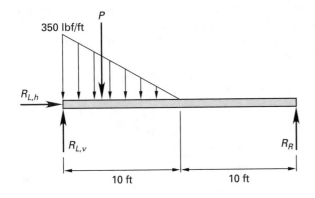

The equivalent point load for the triangular load is located at one-third the length of the load (one-third of 10 ft) and is

$$P = \tfrac{1}{2}bh = \left(\frac{1}{2}\right)(10 \text{ ft})\left(350 \; \frac{\text{lbf}}{\text{ft}}\right)$$
$$= 1750 \text{ lbf}$$

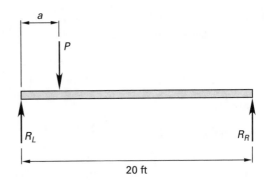

Summing moments about the left support,

$$\Sigma M = 0 \text{ lbf} = Pa - R_R(20 \text{ ft})$$
$$0 \text{ lbf} = (1750 \text{ lbf})\left(\frac{10 \text{ ft}}{3}\right) - R_R(20 \text{ ft})$$
$$R_R = 292 \text{ lbf}$$
$$R_L = P - R_R = 1750 \text{ lbf} - 292 \text{ lbf}$$
$$= 1458 \text{ lbf}$$

The shear diagram for this beam is as follows.

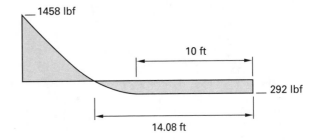

From the shear diagram, determine that the point of zero shear is between 0 ft and 10 ft from the left support. In this case, it is easier to write the equation for shear at a distance greater than 10 ft from the right support.

Shear is equal to the area under the load. The area under the load at a distance x from the base of the triangle is

$$A_x = \tfrac{1}{2}x\left(\frac{350 \; \frac{\text{lbf}}{\text{ft}}}{10 \text{ ft}}\right)x$$

The equation for shear at a distance x measured from greater than 10 ft from the right support is

$$V_x = R_R - A_x = 292 \text{ lbf} - \tfrac{1}{2}x\left(\frac{350 \; \frac{\text{lbf}}{\text{ft}}}{10 \text{ ft}}\right)x$$
$$= 292 \text{ lbf} - \left(\frac{1}{2}\right)\left(35 \; \frac{\text{lbf}}{\text{ft}^2}\right)x^2$$

Shear equals 0 lbf at

$$0 \text{ lbf} = 292 \text{ lbf} - \left(17.5 \; \frac{\text{lbf}}{\text{ft}^2}\right)x^2$$
$$x = 4.09 \text{ ft} \quad [\text{measured left of center}]$$

The distance to the point of zero shear, measured from the left end of the beam, is

$$D_{0 \text{ shear},L} = 10 \text{ ft} - x = 10 \text{ ft} - 4.09 \text{ ft}$$
$$= 5.91 \text{ ft} \quad (5.9 \text{ ft})$$

The answer is (C).

Why Other Options Are Wrong

(A) This incorrect solution writes the shear equation at a point measured from the left support but does not include the equivalent point load in the shear equation.

The equation for shear at a distance x from the left support is

$$V_x = R_L - \left(\frac{1}{2}(10 \text{ ft} - x) \left(\frac{350 \frac{\text{lbf}}{\text{ft}}}{10 \text{ ft}} \right) (10 \text{ ft} - x) \right)$$

$$= 1458 \text{ lbf}$$

$$- \left(\left(\frac{1}{2} \right) (10 \text{ ft} - x) \left(\frac{350 \frac{\text{lbf}}{\text{ft}}}{10 \text{ ft}} \right) (10 \text{ ft} - x) \right)$$

$$= 1458 \text{ lbf} - \left(17.5 \frac{\text{lbf}}{\text{ft}^2} \right) (10 \text{ ft} - x)^2$$

Shear equals 0 lbf at

$$0 \text{ lbf} = 1458 \text{ lbf} - \left(17.5 \frac{\text{lbf}}{\text{ft}^2} \right) (10 \text{ ft} - x)^2$$

$$(10 \text{ ft} - x)^2 = 83.31 \text{ ft}^2$$

$$x = 0.87 \text{ ft}$$

(B) This incorrect solution calculates the distance not from the left support, but rather from the center of the beam.

(D) This incorrect solution distributes the load over the entire beam instead of just one-half of it.

SOLUTION 6

Limit eccentricity, e, to one-sixth of the footing length to keep the soil resultant within the kern limit and avoid tension in the soil.

$$e = \frac{M}{P} = \frac{M_{\text{wind}}}{P_D + P_L}$$

$$= \frac{240 \text{ ft-kips}}{80 \text{ kips} + 100 \text{ kips}}$$

$$= 1.33 \text{ ft}$$

$$e \leq \frac{L}{6}$$

$$1.33 \text{ ft} \leq \frac{L}{6}$$

Solving for L,

$$L \geq (1.33 \text{ ft})(6) = 7.98 \text{ ft} \quad (8.0 \text{ ft})$$

If completing a footing design, the allowable soil press
The answer is (B). *would be checked & footing size*
                      adjusted accordingly.

Why Other Options Are Wrong

(A) This incorrect solution calculates the required size of a square footing (ignoring the footing width given) based on the dead and live loads and allowable soil pressure. The problem statement is asking for the minimum length required for the entire footing to be effective, not the minimum size based on the soil pressure.

(C) This incorrect solution calculates eccentricity using factored loads. Factored loads are used in concrete design, but not for determining footing size. Footing size should be based on unfactored soil pressure.

(D) This incorrect solution calculates the required length based on the dead and live loads and allowable soil pressure. The problem statement is asking for the minimum length required for the entire footing to be effective, not the minimum size based on the soil pressure.

MECHANICS OF MATERIALS

SOLUTION 7 *(Include beam self weight)*
              Answer changes to 0.27 in

Deflection is a function of the moment of inertia of the beam. Using the table of properties found in the AISC ASD manual, for a W16 × 31 beam,

$$I = 375 \text{ in}^4$$
$$E = 29 \times 10^6 \text{ lbf/in}^2$$

From the AISC ASD manual's table Beam Diagrams and Deflections, find the three-span uniformly loaded beam in Loading Diagram 36. The maximum deflection in the first span is given as

$$\Delta_{\text{max}} = \frac{0.0069 w L^4}{EI}$$

$$w = w_D + w_L = 500 \frac{\text{lbf}}{\text{ft}} + 1000 \frac{\text{lbf}}{\text{ft}}$$

$$= 1500 \text{ lbf/ft}$$

$$L = 20 \text{ ft}$$

$$\Delta_{\text{max}} = \frac{0.0069 w L^4}{EI}$$

$$= \frac{(0.0069) \left(1500 \frac{\text{lbf}}{\text{ft}} \right) (20 \text{ ft})^4 \left(1728 \frac{\text{in}^3}{\text{ft}^3} \right)}{\left(29 \times 10^6 \frac{\text{lbf}}{\text{in}^2} \right) (375 \text{ in}^4)}$$

$$= 0.26 \text{ in}$$

The answer is (B).

Why Other Options Are Wrong

(A) This incorrect solution approximates the deflection of the first span by using Loading Diagram 12, for a beam pinned at one end and fixed at the other, from the AISC ASD manual's table Beam Diagrams and Deflections. Although a continuous-span end condition is sometimes approximated by a fixed end, the actual loading diagram should be used when available.

(C) When calculating the deflection, this incorrect solution uses the moment of inertia for a W16 × 26 beam instead of a W16×31 beam from the AISC ASD manual beam properties table.

(D) This incorrect solution does not consider that the spans are continuous. It finds the maximum deflection for a uniformly loaded beam with pinned supports, using Loading Diagram 1 from the AISC ASD manual's table Beam Diagrams and Deflections.

SOLUTION 8

Using the moment-area method, the deflection of a beam at a particular point is equal to the moment of the M/EI diagram about that point.

The moment at the fixed end of the beam shown is

$$M = Pl = (10 \text{ kips})(25 \text{ ft})$$
$$= 250 \text{ ft-kips}$$

The moment diagram is

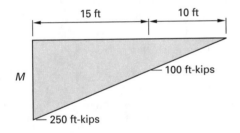

To draw the M/EI diagram, divide the moment diagram by the respective moment of inertia.

$$\frac{M_{\text{support}}}{EI} = \frac{250 \text{ ft-kips}}{E(2000 \text{ in}^4)}$$
$$= \frac{0.125 \text{ ft-kips}}{E \text{ in}^4}$$

The M/EI diagram is

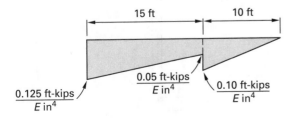

The deflection at the free end of the beam is the moment of the M/EI diagram about the free end.

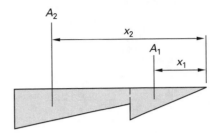

$$\delta = A_1 x_1 + A_2 x_2$$
$$= \left(\frac{1}{2}\right)(10 \text{ ft})\left(\frac{0.10 \text{ ft-kips}}{E \text{ in}^4}\right)\left(\frac{2}{3}\right)(10 \text{ ft})$$
$$+ \left(\frac{0.05 \text{ ft-kips}}{E \text{ in}^4}\right)(15 \text{ ft})\left(10 \text{ ft} + \frac{15 \text{ ft}}{2}\right)$$
$$+ \left(\frac{1}{2}\right)(15 \text{ ft})\left(\frac{0.125 \text{ ft-kips} - 0.05 \text{ ft-kips}}{E \text{ in}^4}\right)$$
$$\times \left((15 \text{ ft})\left(\frac{2}{3}\right) + 10 \text{ ft}\right)$$
$$= \frac{27.7 \frac{\text{ft}^3\text{-kips}}{\text{in}^4}}{E}$$
$$= \left(\frac{27.7 \frac{\text{ft}^3\text{-kips}}{\text{in}^4}}{29 \times 10^6 \frac{\text{lbf}}{\text{in}^2}}\right)\left(1728 \frac{\text{in}^3}{\text{ft}^3}\right)\left(1000 \frac{\text{lbf}}{\text{kip}}\right)$$
$$= 1.65 \text{ in} \quad (1.7 \text{ in})$$

The answer is (B).

Why Other Options Are Wrong

(A) This incorrect solution does not convert kips to pounds in calculating the deflection. The units do not work out.

(C) This incorrect solution reverses I_1 and I_2 when drawing the M/EI diagram.

(D) This incorrect solution does not make the adjustment for the differing moment of inertia over the span length. The deflection is incorrectly calculated using a moment of inertia of 1000 in^4 for the entire length of the beam.

SOLUTION 9

To find the centroid of the cross section, divide the section into two rectangles.

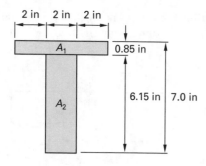

$$A_1 = (0.85 \text{ in})(6 \text{ in})$$
$$= 5.1 \text{ in}^2$$
$$A_2 = (6.15 \text{ in})(2 \text{ in})$$
$$= 12.3 \text{ in}^2$$

The distance from the top of the section to the centroid is

$$y_c = \frac{\sum A_i y_{ci}}{\sum A_i}$$

$$= \frac{(5.1 \text{ in}^2)\left(\dfrac{0.85 \text{ in}}{2}\right) + (12.3 \text{ in}^2)\left(\dfrac{6.15 \text{ in}}{2} + 0.85 \text{ in}\right)}{5.10 \text{ in}^2 + 12.3 \text{ in}^2}$$

$$= 2.9 \text{ in} \quad [\text{from top of section}]$$

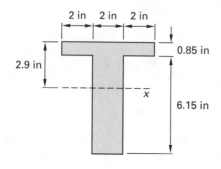

The shear stress is

$$\tau = \frac{VQ}{Ib}$$

Calculate I and Q. For a rectangular section,

$$I = \frac{bh^3}{12}$$

$$I_1 = \frac{(6 \text{ in})(0.85 \text{ in})^3}{12}$$
$$= 0.31 \text{ in}^4$$

$$I_2 = \frac{(2 \text{ in})(6.15 \text{ in})^3}{12}$$
$$= 38.8 \text{ in}^4$$

The moment of inertia about the centroid is

$$I_x = \sum (I_i + A_i d_i^2)$$

$$= 0.31 \text{ in}^4 + (5.1 \text{ in}^2)\left(2.9 \text{ in} - \frac{0.85 \text{ in}}{2}\right)^2$$
$$+ 38.8 \text{ in}^4$$
$$+ (12.3 \text{ in}^2)\left(7.0 \text{ in} - \frac{6.15 \text{ in}}{2} - 2.9 \text{ in}\right)^2$$

$$= 83.3 \text{ in}^4$$

The statical moment of the area, Q, is the product of the area above or below the point in question and the distance from the centroidal axis to the centroid of the area. Looking at the area below the centroid,

$$Q = A\bar{y} = \tfrac{1}{2}b(h - y_c)^2$$
$$= \left(\frac{1}{2}\right)(2 \text{ in})(7.0 \text{ in} - 2.9 \text{ in})^2$$
$$= 16.8 \text{ in}^3$$

The shear stress at the centroid is

$$\tau = \frac{VQ}{Ib} = \frac{(100 \text{ lbf})(16.8 \text{ in}^3)}{(83.3 \text{ in}^4)(2 \text{ in})}$$
$$= 10.1 \text{ lbf/in}^2 \quad (10 \text{ lbf/in}^2)$$

The answer is (C).

Why Other Options Are Wrong

(A) This incorrect solution miscalculates the location of the centroid of the section. The distance from the centroid of area A_2 to the top of the section fails to include the 0.85 in thickness of area A_1.

(B) This incorrect solution miscalculates A_2 and carries the mistake throughout the subsequent calculations. The length of A_2 is taken as the overall length of 7.0 in instead of the actual length of 6.15 in.

(D) This incorrect solution does not properly calculate the transformed moment of inertia. It directly adds the moments of inertia for each area instead of calculating the transformed moment of inertia.

SOLUTION 10

Find the total area of the section.

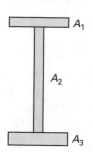

$$A = A_1 + A_2 + A_3 = (0.75 \text{ in})(4.75 \text{ in})$$
$$+ (10 \text{ in} - 0.75 \text{ in} - 1.0 \text{ in})(0.75 \text{ in})$$
$$+ (1.0 \text{ in})(4.75 \text{ in})$$
$$= 14.5 \text{ in}^2$$

Compressive stress is

$$\sigma = \frac{P}{A} = \frac{560 \text{ lbf}}{14.5 \text{ in}^2}$$
$$= 38.6 \text{ lbf/in}^2 \quad (39 \text{ lbf/in}^2)$$

The answer is (C).

Why Other Options Are Wrong

(A) This incorrect solution calculates the section area, not the stress.

(B) This incorrect solution miscalculates area A_2 by using the overall length of the section (10 in) for the length of A_2.

(D) This incorrect solution miscalculates A_1 and A_3 by using 4.0 in for the width of both sections instead of 4.75 in.

MATERIALS

SOLUTION 11

ACI 318 specifies that concrete exposed to freezing and thawing conditions must be air-entrained for durability (see ACI 318 Sec. 4.2). The total air content for frost-resistant concrete can be found in ACI 318 Table 4.2.1. The commentary for this section indicates that pavements, sidewalks, and parking garages are all examples of applications that experience severe exposure. For a maximum aggregate size of 1 in and severe exposure conditions, the total air content of the concrete mix should be 6%.

The answer is (D).

Why Other Options Are Wrong

(A) In this incorrect solution, the units on the table have been misread and the value is "converted" to a percentage by dividing by 100.

(B) This incorrect solution uses the maximum water-cementitious materials ratio from ACI 318 Table 4.2.2 (Requirements for Special Exposure Conditions) for concrete exposed to freezing and thawing instead of the ratio from ACI 318 Table 4.2.1 (Total Air Content for Frost-Resistant Concrete).

(C) This is the value from ACI 318 Table 4.2.1 for moderate exposure, not severe exposure.

SOLUTION 12

Chapter 13 of ACI 318 covers two-way slab systems. Section 13.6 covers the Direct Design Method. Statement I is true.

Section 13.6.1.1 of ACI 318 requires that there be at least three continuous spans in each direction in order to justify use of the Direct Design Method. Statement II is false.

Section 13.7 of ACI 318 gives the requirements for the Equivalent Frame Method. There are no limitations in this section on the number and length of spans that can be analyzed using this method. Statement III is true.

Section 13.6.1.5 of ACI 318 limits the live load to two times the dead load for slabs using the Direct Design Method. Statement IV is true.

The answer is (C).

Why Other Options Are Wrong

(A) This answer is incorrect. Although statement I is true, statement II is false.

(B) This answer is incorrect. Although statements I and III are true, so is statement IV.

(D) This answer is incorrect. Although statements I, III, and IV are true, statement II is false.

SOLUTION 13

The shear capacity of a fillet weld, R_v, does not depend on whether the weld is horizontal or vertical but on the properties of the materials joined and the size and length of the weld. The shear capacity of a fillet weld can be determined from Table J2.5 of the AISC ASD specification.

$$R_v = 0.30 F_{u,\text{rod}} A_e$$
$$= 0.30 F_{u,\text{rod}} L_w t_e$$

Since t_e is the effective throat thickness, statement I is true.

Since L_w is the length of the weld, statement II is true.

Since $F_{u,\text{rod}}$ is the tensile strength of the weld metal, statement IV is true.

The answer is (D).

Why Other Options Are Wrong

(A) This incorrect solution selects the only statement that is false.

(B) This incorrect solution does not include statement IV, which is true.

(C) This incorrect solution includes statement III, which is false. Whether the weld is horizontal or vertical does not affect the shear capacity of the weld.

SOLUTION 14

Use the Column Base Plates Design Procedure given in AISC ASD Ch. 3.

For a W12 × 72 steel column,

$$d = 12.25 \text{ in}$$
$$b_f = 12.04 \text{ in}$$

From the AISC ASD specification Sec. J9, the allowable bearing pressure on the concrete spread footing when less than the full area of the footing is covered is

$$F_p = 0.35 f_c' \sqrt{\frac{A_2}{A_1}} \leq 0.7 f_c'$$

$$A_1 = BN = (14 \text{ in})(16 \text{ in}) = 224 \text{ in}^2$$
$$\left[\begin{array}{c} \text{See Fig. 1 of the AISC ASD Column} \\ \text{Base Plates Design Procedure.} \end{array} \right]$$

$$A_2 = bL = (8 \text{ ft})\left(12 \frac{\text{in}}{\text{ft}}\right)(8 \text{ ft})\left(12 \frac{\text{in}}{\text{ft}}\right)$$
$$= 9216 \text{ in}^2$$

$$F_{p1} = 0.35 f_c' \sqrt{\frac{A_2}{A_1}} = (0.35)\left(3000 \frac{\text{lbf}}{\text{in}^2}\right)\sqrt{\frac{9216 \text{ in}^2}{224 \text{ in}^2}}$$
$$= 6735 \text{ lbf/in}^2$$

$$F_{p2} = 0.7 f_c' = (0.7)\left(3000 \frac{\text{lbf}}{\text{in}^2}\right)$$
$$= 2100 \text{ lbf/in}^2$$

The allowable bearing pressure is the smaller of F_{p1} or F_{p2}, which is 2100 lbf/in².

The actual bearing pressure is

$$f_p = \frac{P}{A_1} = \frac{420 \text{ kips}}{224 \text{ in}^2}$$
$$= 1.875 \text{ kips/in}^2$$

The thickness of the plate is

$$t_p = 2c\sqrt{\frac{f_p}{F_y}}$$

c is the larger of m, n, and $\lambda n'$.

$$m = \frac{N - 0.95d}{2} = \frac{16 \text{ in} - (0.95)(12.25 \text{ in})}{2}$$
$$= 2.18 \text{ in}$$
$$n = \frac{B - 0.80 b_f}{2} = \frac{14 \text{ in} - (0.80)(12.04 \text{ in})}{2}$$
$$= 2.18 \text{ in}$$

$$\lambda = \frac{2(1 - \sqrt{1 - q})}{\sqrt{q}} \leq 1.0$$
$$q = \frac{4 f_p d b_f}{(d + b_f)^2 F_p} < 1.0$$
$$n' = \frac{\sqrt{d b_f}}{4}$$

Conservatively, set λ equal to 1.0 per the AISC ASD Column Base Plates Design Procedure.

$$\lambda n' = 1.0 \left(\frac{\sqrt{d b_f}}{4} \right)$$
$$= (1.0)\left(\frac{\sqrt{(12.25 \text{ in})(12.04 \text{ in})}}{4} \right)$$
$$= 3.04 \text{ in}$$
$$c = 3.04 \text{ in}$$

Alternately, use $\lambda = \left(2(1 - \sqrt{1 - q})\right)/\sqrt{q} \leq 1.0$.

$$q = \frac{(4)\left(1.875 \frac{\text{kips}}{\text{in}^2}\right)(12.25 \text{ in})(12.04 \text{ in})\left(1000 \frac{\text{lbf}}{\text{kip}}\right)}{(12.25 \text{ in} + 12.04 \text{ in})^2 \left(2100 \frac{\text{lbf}}{\text{in}^2}\right)}$$
$$= 0.893$$

$$\lambda = \frac{(2)\left(1 - \sqrt{1 - 0.893}\right)}{\sqrt{0.893}}$$
$$= 1.42 \leq 1.0$$
$$\lambda = 1.0$$

The thickness of the plate is

$$t_p = (2)(3.04 \text{ in})\sqrt{\frac{1.875 \frac{\text{kips}}{\text{in}^2}}{36 \frac{\text{kips}}{\text{in}^2}}}$$
$$= 1.39 \text{ in}$$

The answer is (B).

Why Other Options Are Wrong

(A) This incorrect solution neglects to calculate $\lambda n'$, which in this case is the controlling distance.

(C) This incorrect solution reverses N and B when calculating m and n. N is the length of the base plate in the direction of the column depth. B is the width of the base plate in the direction of the column flange, as shown in Fig. 1 of the AISC ASD Column Base Plates Design Procedure.

(D) This incorrect solution uses the compressive strength of concrete, f_c', for bearing pressure, f_p. See the AISC ASD specification Sec. J9 or ACI 318 for the proper bearing pressure equation.

SOLUTION 15

Section 2.3 and Table 2.3.1 of the NDS contain the adjustment factors and their applicability for design values.

The volume factor, C_V, is discussed in the footnotes to NDS Table 2.3.1. Footnote 3 states, "The volume factor, C_V, shall apply only to glued laminated timber bending members (see Sec. 5.3.2)." Statement II is true.

The load duration factor, C_D, is discussed in NDS Sec. 2.3.2 and NDS Table 2.3.2. Both NDS Sec. 2.3.2.1 and Footnote 1 to NDS Table 2.3.2 state that "Load duration factors shall not apply to modulus of elasticity, E..." Statement IV is true.

The answer is (C).

Why Other Options Are Wrong

(A) Although statement II is true, statement I is false. The temperature factor, C_t, is discussed in NDS Sec. 2.3.4. The temperature factor applies to members subjected to sustained exposure to elevated (over 100°F) temperatures. Because the statement applies to members subjected to extreme cold, not heat, statement I is false.

(B) Although statement II is true, statement III is false. The repetitive member factor, C_r, is discussed in Footnote 5 of NDS Table 2.3.1. Footnote 5 states, "The repetitive member factor, C_r, shall apply only to dimension lumber bending members 2 in to 4 in thick (see Sec. 4.3.4)." Because the joists in this case are 1 in thick, statement III is false.

(D) Although statement IV is true, statement III is false because the repetitive member factor, C_r, applies only to dimension lumber bending members 2 in to 4 in thick. The joists in this case are 1 in thick.

MEMBER DESIGN

SOLUTION 16

The weight of brick veneer is given as 40 lbf/ft^2. Brick veneer is typically supported at each floor. Therefore, the brick load on spandrel beam BC is for a one-story height, or 13 ft. The spacing between beams is 10 ft. The tributary width, w', for a spandrel beam is

$$w' = \left(\frac{1}{2}\right)(10 \text{ ft}) = 5 \text{ ft}$$

The load on the spandrel beam is

$$w_{\text{per ft}} = w' w_{\text{per ft}^2}$$

$$w_D = (5 \text{ ft})\left(60 \; \frac{\text{lbf}}{\text{ft}^2}\right) = 300 \text{ lbf/ft}$$

$$w_L = (5 \text{ ft})\left(40 \; \frac{\text{lbf}}{\text{ft}^2}\right) = 200 \text{ lbf/ft}$$

$$w_{\text{brick}} = (13 \text{ ft})\left(40 \; \frac{\text{lbf}}{\text{ft}^2}\right) = 520 \text{ lbf/ft}$$

$$\begin{aligned} w_{\text{total}} &= w_D + w_L + w_{\text{brick}} \\ &= 300 \; \frac{\text{lbf}}{\text{ft}} + 200 \; \frac{\text{lbf}}{\text{ft}} + 520 \; \frac{\text{lbf}}{\text{ft}} \\ &= 1020 \text{ lbf/ft} \end{aligned}$$

The maximum moment on the beam is

$$\begin{aligned} M_{\text{max}} &= \frac{w_{\text{total}}L^2}{8} = \frac{\left(1020 \; \frac{\text{lbf}}{\text{ft}}\right)(30 \text{ ft})^2}{8} \\ &= 114{,}750 \text{ ft-lbf} \quad (115 \text{ ft-kips}) \end{aligned}$$

Using the AISC ASD manual's Selection Table, the lightest beam required based on maximum moment is a W16 × 40 (in boldface type) ($M_R = 128$ ft-kips).

Check for deflection. The maximum deflection of the beam cannot exceed $L/600$ or 0.3 in, whichever is less.

$$\frac{L}{600} = \frac{(30 \text{ ft})\left(12 \; \frac{\text{in}}{\text{ft}}\right)}{600} = 0.6 \text{ in}$$

Therefore, the deflection cannot exceed 0.3 in.

Deflection for a simply supported beam is

$$\Delta = \frac{5w_{\text{total}}L^4}{384EI} \le 0.3 \text{ in}$$

Solving for I gives

$$\begin{aligned} I &\ge \frac{5w_{\text{total}}L^4}{384E\Delta} \\ &\ge \frac{(5)\left(1020 \; \frac{\text{lbf}}{\text{ft}}\right)(30 \text{ ft})^4\left(1728 \; \frac{\text{in}^3}{\text{ft}^3}\right)}{(384)\left(29 \times 10^6 \; \frac{\text{lbf}}{\text{in}^2}\right)(0.3 \text{ in})} \\ &\ge 2137 \text{ in}^4 \end{aligned}$$

Using the AISC ASD manual's Moment of Inertia Selection Table, the lightest beam required based on deflection is a W24 × 84 (in boldface type) ($I_x = 2370$ in^4). Deflection controls. Use a W24 × 84 beam.

Check the additional deflection due to the self-weight of the beam.

$$\begin{aligned} \Delta &= \frac{5w_{\text{total}}L^4}{384EI} \\ &= \frac{(5)\left(84 \; \frac{\text{lbf}}{\text{ft}} + 1020 \; \frac{\text{lbf}}{\text{ft}}\right)(30 \text{ ft})^4\left(1728 \; \frac{\text{in}^3}{\text{ft}^3}\right)}{(384)\left(29 \times 10^6 \; \frac{\text{lbf}}{\text{in}^2}\right)(2370 \text{ in}^4)} \\ &= 0.29 \text{ in} < 0.3 \text{ in} \quad [\text{OK}] \end{aligned}$$

For relatively short spans (generally 15 ft or less for all but the largest sections), shear stress in the web should also be checked. In this case the span is 30 ft, so bending stress controls.

The answer is (D).

Why Other Options Are Wrong

(A) This incorrect solution neglects the weight of the facade when calculating the load on the spandrel beam and ignores the deflection limitation of $L/600$ or 0.3 in.

(B) This incorrect solution correctly calculates the moment on the beam but ignores the deflection limitation of $L/600$ or 0.3 in.

(C) This incorrect solution correctly calculates the moment on the beam but uses the wrong deflection limitation. The deflection is limited to $L/600$ or 0.3 in, whichever is less. In this case, $L/600$ is greater than 0.3 in, so the deflection should be limited to 0.3 in.

Note that in this case, the beam size remains the same if the weight of the brick is neglected and the moment and deflection are otherwise correctly determined.

SOLUTION 17

The values in the columns tables in the AISC ASD manual are based on an effective length with respect to the minor axis, KL_y.

The effective column lengths are given as

$$KL_x = 32 \text{ ft}$$
$$KL_y = 18 \text{ ft}$$

Enter the AISC ASD columns tables with an effective length of 18 ft. Since the column depth is limited to 12 in, begin with the W12 sections. For a yield stress of 50 kips/in^2, use the shaded columns and determine that a W12 × 87 column has a capacity of 534 kips. Check the capacity based on the effective length for buckling about the x-axis.

From the AISC ASD columns tables, $r_x/r_y = 1.75$.

The equivalent effective length for the x-axis is

$$\frac{32 \text{ ft}}{1.75} = 18.3 \text{ ft} > 18 \text{ ft}$$

Therefore, check the capacity for $KL = 18.3$ ft.

By interpolation, a W12 × 87 column has a capacity of 528 kips [OK].

The answer is (A).

Why Other Options Are Wrong

(B) This solution incorrectly uses the column for A36 steel in the AISC ASD columns tables to find that a W12 × 120 column is needed.

(C) This incorrect solution reverses the major and minor axes in determining effective length.

(D) This incorrect solution reverses the major and minor axes in determining effective length and uses the column for Grade 36 steel instead of Grade 50 steel in the AISC ASD columns tables.

SOLUTION 18

Section 10.3.6 of ACI 318 gives the design axial load strength for compression members. Solve ACI 318 Eq. 10-1 for the gross area of the concrete.

$$\rho_{g_{max}} = 0.08 \quad [\text{ACI 318 Sec. 10.9.1}]$$

$$\beta = 0.85 \quad \begin{bmatrix} \text{for spiral columns,} \\ \text{ACI 318 Sec. 10.3.5.3} \end{bmatrix}$$

concrete strengths upto 4000 lbf/i   *10.2.7.3*

$$\phi = 0.75 \quad \begin{bmatrix} \text{for spiral columns,} \\ \text{ACI 318 Sec. 10.3.5.3} \end{bmatrix}$$

0.70   *9.3.2.2*

$$P_u = 1.4D + 1.7L$$

1.2   *1.6*

$$= (1.4)(300 \text{ kips}) + (1.7)(350 \text{ kips})$$

1.2   *1.6*

$$= 1015 \text{ kips}$$

920

$$A_g = \frac{P_u}{\phi\beta\big(0.85f_c'(1-\rho_g) + \rho_g f_y\big)}$$

920

$$= \frac{(1015 \text{ kips})\left(1000 \dfrac{\text{lbf}}{\text{kip}}\right)}{(0.75)(0.85)\left(\begin{array}{l}(0.85)\left(4000 \dfrac{\text{lbf}}{\text{in}^2}\right)(1-0.08) \\ + (0.08)\left(60{,}000 \dfrac{\text{lbf}}{\text{in}^2}\right)\end{array}\right)}$$

0.70   *920*

$$= \frac{1{,}015{,}000 \text{ lbf}}{(0.75)(0.85)\left(7928 \dfrac{\text{lbf}}{\text{in}^2}\right)}$$

$$= 201 \text{ in}^2 \quad (200 \text{ in}^2)$$

195

The answer is (C).

Why Other Options Are Wrong

(B) (A) This incorrect solution does not apply the load factors to calculate the factored load but uses unfactored loads instead. *Uses load factors from ACI-318-99*

(A) (B) This incorrect solution makes a mathematical error in calculating the cross-sectional area by not multiplying the entire denominator by $\phi\beta$.

$$A_g = \frac{P_u}{\phi\beta\left(0.85f'_c(1-\rho_g)+\rho_g f_y\right)}$$

$$= \frac{(\overset{920}{\cancel{1015 \text{ kips}}})\left(1000 \ \frac{\text{lbf}}{\text{kip}}\right)}{(\cancel{0.75})(0.85)(0.85)\left(4000 \ \frac{\text{lbf}}{\text{in}^2}\right)}$$

0.70

$$\times (1-0.08)+(0.08)\left(60,000 \ \frac{\text{lbf}}{\text{in}^2}\right)$$

$$= \frac{1{,}015{,}000 \text{ lbf}}{1994 \ \frac{\text{lbf}}{\text{in}^2} + 4800 \ \frac{\text{lbf}}{\text{in}^2}}$$

$$= \cancel{149 \text{ in}^2} \quad (\cancel{150 \text{ in}^2})$$

135   140

(D) This incorrect solution uses the adjustment factors for a tied column ($\cancel{\beta=0.80}$ and $\phi = \cancel{0.70}$) rather than for a spiral column.

0.65

SOLUTION 19

The minimum thickness of a footing is determined by the greater of the minimum depth required for shear and the minimum depth required by Sec. 15 of ACI 318.

The required depth for shear is based on the ultimate soil pressure. The ultimate soil pressure is

$$q_u = \frac{\overset{1.2}{\cancel{1.4}}P_D + \overset{1.6}{\cancel{1.7}}P_L}{A}$$

$$= \frac{(1.4)(50 \text{ kips})+(1.7)(75 \text{ kips})}{(6.5 \text{ ft})(6.5 \text{ ft})}$$

4.26

$$= \cancel{4.67} \text{ kips/ft}^2$$

From ACI 318 Sec. 9.3.2, determine that the strength reduction factor, ϕ, for shear is 0.85.

From Ch. 11 of ACI 318, the allowable shear in the concrete footing, assuming two-way action, is

$$v_c = 4\phi\sqrt{f'_c}$$

$$= (4)(\overset{0.75}{\cancel{0.85}})\left(\sqrt{3000} \ \frac{\text{lbf}}{\text{in}^2}\right)\left(\frac{1 \text{ kip}}{1000 \text{ lbf}}\right)\left(144 \ \frac{\text{in}^2}{\text{ft}^2}\right)$$

23.7

$$= \cancel{26.8} \text{ kips/ft}^2$$

✗ Or from a table giving allowable shear as a function of concrete strength, such as Table 8-2 of *Foundation Analysis and Design* by Joseph E. Bowles or a similar reference, find that

$$v_c = 26.8 \text{ kips/ft}^2$$

Sum the shear forces acting on the footing to determine the footing depth. For a concrete footing for a square concrete column, the equation becomes

$$d^2\left(v_c + \frac{q_u}{4}\right)+d\left(v_c + \frac{q_u}{2}\right)w = (B^2 - w^2)\left(\frac{q_u}{4}\right)$$

$$d^2\left(26.8 \ \frac{\text{kips}}{\text{ft}^2} + \frac{4.67 \ \frac{\text{kips}}{\text{ft}^2}}{4}\right)$$

$$+ d\left(26.8 \ \frac{\text{kips}}{\text{ft}^2} + \frac{4.67 \ \frac{\text{kips}}{\text{ft}^2}}{2}\right)(1.0 \text{ ft})$$

$$= \left((6.5 \text{ ft})^2 - (1.0 \text{ ft})^2\right)\left(\frac{4.67 \ \frac{\text{kips}}{\text{ft}^2}}{4}\right)$$

$$d^2\left(27.97 \ \frac{\text{kips}}{\text{ft}^2}\right)+d\left(29.14 \ \frac{\text{kips}}{\text{ft}^2}\right)(1.0 \text{ ft})$$

$$= 48.16 \text{ kips}$$

Using the quadratic equation and solving for d gives

$$d = (\overset{0.91}{\cancel{0.89 \text{ ft}}})\left(12 \ \frac{\text{in}}{\text{ft}}\right)$$

$$= \cancel{10.7 \text{ in}}$$

10.9

The minimum thickness of the footing is

$$h = d + \tfrac{1}{2}\left(d_{b,x} + d_{b,y}\right) + \text{cover}$$

A no. 5 bar has a diameter of 0.625 in (ACI 318 App. E). The minimum cover (below the bars) for concrete against the earth is 3 in (ACI 318 7.7.1).

$$h = 10.7 \text{ in} + \left(\tfrac{1}{2}\right)(0.625 \text{ in} + 0.625 \text{ in})+3 \text{ in}$$

$$= \cancel{14.3 \text{ in}} \quad (\cancel{15 \text{ in}}) \quad \begin{bmatrix} \text{Round up to 15 in because} \\ h \text{ must be } \geq 14.3 \text{ in.} \end{bmatrix}$$

14.5"   15"

Check that this depth exceeds the minimum required by the code. ACI 318 Sec. 15.7 specifies that the depth of a footing above the bottom of reinforcement must be 6 in.

$$d = 10.7 \text{ in} > 6 \text{ in} \quad [\text{OK}]$$

The answer is (D). ✓

Why Other Options Are Wrong

(A) This incorrect solution does not convert the allowable shear stress to kips/ft^2 in the calculation of depth to reinforcement.

(B) This incorrect solution neglects to add cover to determine the overall footing depth.

(C) This incorrect solution uses unfactored loads. Concrete should be designed using ultimate strength equations and factored loads.

SOLUTION 20

From ACI 318 Table 9.5(a), determine the minimum thickness, h, for a solid one-way slab with one end continuous.

$$h = \frac{l}{24}$$

l is the span length in inches.

$$l = (\underset{14}{\cancel{15}}\text{ ft})\left(12\ \frac{\text{in}}{\text{ft}}\right) = \underset{168}{\cancel{180}}\text{ in}$$

$$h = \frac{180\text{ in}}{24} = \underset{7.0}{\cancel{7.5}}\text{ in}$$

For lightweight concrete, the footnotes to the table indicate that the tabulated value must be modified.

$$h' = (1.65 - 0.005w_c)h$$

From ACI 318 Table 9.5(a), w_c is 100. Therefore,

$$h' = \bigl(1.65 - (0.005)(100)\bigr)(7.5\text{ in})$$
$$= \underset{8.1}{\cancel{8.6}}\text{ in}$$

The answer is (C).
(B)

Why Other Options Are Wrong

(A) This incorrect solution neglects to adjust the tabulated value for lightweight concrete. From ACI 318 Table 9.5(a), determine the minimum thickness, h, for a solid one-way slab with one end continuous.

(C) (B) This incorrect solution uses the ~~clear~~ span [c/c span] for l instead of the ~~center-to-center~~ [clear] span. The clear span can only be used for slab spans no more than 10 ft.

(D) This incorrect solution results from assuming that the slab is simply supported when selecting the thickness equation from ACI 318 Table 9.5(a).

Depth Solutions

LOADINGS

SOLUTION 21

Section 1617.5 of the IBC contains the simplified analysis procedure for seismic design of buildings. The seismic base shear is

$$V = \left(\frac{1.2S_{DS}}{R}\right)W \quad \text{[IBC Eq. 16-56]}$$

The design elastic response acceleration at the short period, S_{DS}, is a function of the maximum considered earthquake spectral response accelerations for the short period, S_{MS}, determined using IBC Sec. 1615.1.3.

$$S_{DS} = \tfrac{2}{3}S_{MS}$$
$$= \left(\frac{2}{3}\right)(2.12) = 1.41 \quad \text{[IBC Eq. 16-40]}$$

R is the response modification factor from IBC Table 1617.6.2 and is a function of the basic seismic-force-resisting system. For a bearing wall system, use Part 1 of the table. Based on System H, for ordinary reinforced masonry shear walls, R is 2.5.

$$V = \left(\frac{(1.2)(1.41)}{2.5}\right)W$$
$$= 0.68W$$

The answer is (C).

Why Other Options Are Wrong

(A) This incorrect solution finds the response modification factor in IBC Table 1817.6.2 based on System C, which gives an R of 4.5 for ordinary reinforced concrete shear walls.

(B) This incorrect solution finds the response modification factor in IBC Table 1817.6.2 based on System C, which gives an R of 4.5 for ordinary reinforced concrete shear walls, and assumes the design elastic response acceleration at the short period, S_{DS}, to be given as 2.12.

(D) This incorrect solution assumes the design elastic response acceleration at the short period, S_{DS}, to be given as 2.12.

SOLUTION 22

Section 1617.5 of the IBC defines seismic base shear as directly proportional to effective seismic weight, W. The effective seismic weight is the total dead load plus portions of other loads as listed in the IBC. Therefore, as the building dead load increases, so does the seismic base shear. Statement I is true.

The IBC states the following about the other loads included in the total dead load: "4. 20 percent of flat roof snow load where flat snow load exceeds 30 psf." Statement III is also true.

The answer is (B).

Why Other Options Are Wrong

(A) In areas of high seismic activity, buildings perform best when they have a regular floor plan to more uniformly distribute the seismic loads. Irregular plans often have greater building eccentricities and greater differential seismic movements that are harder to accommodate in the building design. Statement II is false.

(C) In the Northridge earthquake and others, buildings with flexible (or soft) first stories and heavy roofs often failed when the flexible first story swayed beyond the design limits and collapsed. In general, it is preferable to have the more flexible story above a stiffer one. Statement IV is false.

(D) Even though statements I and III are true, statement II is false.

SOLUTION 23

The induced axial load in a constrained member due to a temperature change is

$$P_{th} = \alpha(T_2 - T_1)AE$$

α is the coefficient of thermal expansion of the member. According to Part 6 of the AISC ASD manual, the coefficient of thermal expansion for mild steel is

$$\alpha = \frac{0.00065}{100°\text{F}}$$

SIX-MINUTE SOLUTIONS FOR CIVIL PE EXAM STRUCTURAL PROBLEMS

The area of the steel bar is

$$A = \pi \left(\frac{d^2}{4} \right) = \pi \left(\frac{(2.5 \text{ in})^2}{4} \right) = 4.91 \text{ in}^2$$

$$P_{\text{th}} = \left(\frac{0.00065}{100°\text{F}} \right) (125°\text{F} - 60°\text{F})(4.91 \text{ in}^2)$$

$$\times \left(29 \times 10^6 \; \frac{\text{lbf}}{\text{in}^2} \right) \left(\frac{1 \text{ kip}}{1000 \text{ lbf}} \right)$$

$$= 60 \text{ kips}$$

The answer is (A).

Why Other Options Are Wrong

(B) This incorrect solution uses the maximum temperature instead of the temperature change to calculate the induced axial load.

(C) This solution incorrectly calculates the area of the bar as πd^2 instead of $\pi d^2/4$.

(D) This incorrect solution reads the coefficient of thermal expansion as having no units instead of as 0.00065 per 100°F. The units do not work out in the axial load equation.

SOLUTION 24

Find the rigidity of the walls. Rigidity is the relative stiffness of the walls and is proportional to the inverse of the shear and flexural deflection of the walls. However, for squat walls with h/L no more than 0.25, it is reasonably accurate to calculate the rigidity based on shear alone. For walls with h/L greater than 0.25 and less than 4, both flexure and shear must be considered. For very tall walls, the contribution from shear deformations is very small, and the rigidity can be based on flexure alone.

Since walls B, C, and E have the shortest length, the critical aspect ratio for these walls is

$$\frac{h}{L} = \frac{9 \text{ ft}}{36 \text{ ft}} = 0.25$$

Therefore the rigidity can be based on shear alone. Assume the walls all have the same modulus of elasticity. Since the walls are all the same height, h_j, the rigidity is proportional to the area of the walls. If the walls were of the same thickness, the shear rigidity is proportional to wall length alone.

$$r_j = \frac{k_j}{\sum k_i}$$

$$k_j = \frac{A_j E_j}{h_j}$$

$$r_j = \frac{k_j}{k_j + k_k} = \frac{A_j E_j}{h_j \left(\frac{A_j E_j}{h_j} + \frac{A_k E_k}{h_k} \right)}$$

$$r_j \propto A = tL$$

$$r_A \propto (16 \text{ in}) \left(\frac{1 \text{ ft}}{12 \text{ in}} \right) (60 \text{ ft}) = 80 \text{ ft}^2$$

$$r_B \propto (16 \text{ in}) \left(\frac{1 \text{ ft}}{12 \text{ in}} \right) (36 \text{ ft}) = 48 \text{ ft}^2$$

$$r_C \propto (12 \text{ in}) \left(\frac{1 \text{ ft}}{12 \text{ in}} \right) (40 \text{ ft}) = 40 \text{ ft}^2$$

$$r_D \propto (12 \text{ in}) \left(\frac{1 \text{ ft}}{12 \text{ in}} \right) (36 \text{ ft}) = 36 \text{ ft}^2$$

$$r_E \propto (16 \text{ in}) \left(\frac{1 \text{ ft}}{12 \text{ in}} \right) (36 \text{ ft}) = 48 \text{ ft}^2$$

Find the center of rigidity in each direction.

The distance to the center of rigidity from the South end is

$$\bar{y} = \frac{\sum r_i y_i}{\sum r_i} = \frac{r_A y_A + r_D y_D}{r_A + r_D}$$

$$= \frac{(80 \text{ ft}^2)(160 \text{ ft}) + (36 \text{ ft}^2)(40 \text{ ft})}{80 \text{ ft}^2 + 36 \text{ ft}^2}$$

$$= 123 \text{ ft}$$

The distance to the center of rigidity from the West end is

$$\bar{x} = \frac{\sum r_i x_i}{\sum r_i} = \frac{r_B x_B + r_C x_C + r_E x_E}{r_B + r_C + r_E}$$

$$= \frac{(48 \text{ ft}^2)(80 \text{ ft}) + (40 \text{ ft}^2)(50 \text{ ft}) + (48 \text{ ft}^2)(20 \text{ ft})}{48 \text{ ft}^2 + 40 \text{ ft}^2 + 48 \text{ ft}^2}$$

$$= 50.0 \text{ ft}$$

The answer is (D).

Why Other Options Are Wrong

(A) This incorrect solution does not include the effects of the different wall thickness when calculating the rigidities.

(B) This incorrect solution does not include the effects of the different wall thickness and makes a math error in calculating the distance to the center of rigidity from the West end by calculating $\sum r_i$ equal to 102 ft instead of 112 ft.

(C) This incorrect solution correctly finds the center of rigidity but reverses the direction from which the distance is measured when choosing the answer.

SOLUTION 25

A roof framed with steel joists and corrugated steel deck is typically considered a flexible diagram because of its light weight and flexibility. A rigid diaphragm can be achieved if 2–3 in of concrete is used on the steel decking.

PROFESSIONAL PUBLICATIONS, INC.

For a flexible diaphragm, the lateral loads are distributed according to tributary area. If the roof were a rigid diaphragm (i.e., concrete), the lateral loads would be distributed according to rigidity.

In this case, the shear load can be distributed according to the tributary width, w.

$$V_A = \left(\frac{w_A}{\sum w_i}\right) V_{total}$$

$$= \left(\frac{15 \text{ ft} + \left(\frac{1}{2}\right)(20 \text{ ft})}{85 \text{ ft}}\right)(600 \text{ kips})$$

$$= 176 \text{ kips}$$

The answer is (B).

Why Other Options Are Wrong

(A) This incorrect solution calculates the tributary width of wall A using one-half the distance to the edge of the building instead of the full tributary width.

(C) This incorrect solution distributes the lateral load based on relative rigidities. Relative rigidities can only be used if the diaphragm is rigid. In addition, when loads are based on rigidities, eccentricity of the load (torsion) must also be considered.

(D) This incorrect solution calculates the tributary width of wall A using the full distance between walls instead of one-half the distance between the walls.

SOLUTION 26

Section 1608 of the IBC covers snow loads and references ASCE 7 for most of the design equations. IBC Sec. 1608.4 states that sloped roof snow loads shall be determined in accordance with ASCE 7 Sec. 7.4. Begin by determining the flat roof snow load and adjusting it for the roof slope.

The flat roof snow load is

$$p_f = 0.7 C_e C_t I_s p_g \quad \text{[ASCE 7 Eq. 7-1]}$$

The snow exposure factor, C_e, is given in IBC Table 1608.3.1 based on the terrain exposure category and roof exposure category. The terrain exposure category given in IBC Sec. 1609.4 for an office park is Exposure B. The roof exposure category for a suburban location can be assumed to be partially exposed. Based on these factors, C_e is 1.0.

The thermal factor, C_t, is given in IBC Table 1608.3.2 as 1.1 for ventilated attics with insulation greater than R-25.

The snow load importance factor, I_s, is given in IBC Table 1604.5. An office building falls into Category II. I_s is 1.0.

$$p_f = (0.7)(1.0)(1.1)(1.0)\left(40 \frac{\text{lbf}}{\text{ft}^2}\right) = 30.8 \text{ lbf/ft}^2$$

Check the minimum flat roof snow load as specified in ASCE 7 Sec. 7.3. For ground snow loads over 20 lbf/ft²,

$$p_{min} = I_s\left(20 \frac{\text{lbf}}{\text{ft}^2}\right) = (1.0)\left(20 \frac{\text{lbf}}{\text{ft}^2}\right)$$

$$= 20 \text{ lbf/ft}^2 < 30.8 \text{ lbf/ft}^2 \quad \text{[OK]}$$

IBC Sec. 1608.4 refers to ASCE 7 Sec. 7.4 to determine the sloped roof snow load, p_s.

$$p_s = C_s p_f \quad \text{[ASCE 7 Eq. 7-2]}$$

The roof slope factor, C_s, for cold roofs is found in ASCE 7 Sec. 7.4.2. For a C_t of 1.1, C_s is determined from ASCE 7 Fig. 7-2b using the solid line for an asphalt shingle roof (ASCE 7 Sec. 7.4). For a roof pitch of 6:12, determine that

$$C_s = 1.0 \quad \text{[ASCE 7 Fig. 7-2b]}$$

Therefore, the sloped roof snow load is

$$p_s = (1.0)\left(30.8 \frac{\text{lbf}}{\text{ft}^2}\right) = 30.8 \text{ lbf/ft}^2$$

To determine the maximum snow load, the unbalanced condition must also be considered as found in ASCE 7 Sec. 7.6. For hip and gable roofs, W is the horizontal eave-to-ridge distance, or 20 ft. According to ASCE 7 Sec. 7.6.1, if W is 20 ft or less, the unbalanced snow load on the leeward side is

$$\frac{1.5 p_s}{C_e} = \frac{(1.5)\left(30.8 \frac{\text{lbf}}{\text{ft}^2}\right)}{1.0} = 46.2 \text{ lbf/ft}^2 \quad (46 \text{ lbf/ft}^2)$$

The unbalanced snow load on the windward side is zero.

The unbalanced load condition is greater than the balanced load condition. The maximum snow load on the leeward roof is 46 lbf/ft².

The answer is (C).

Why Other Options Are Wrong

(A) This incorrect solution finds the sloped roof snow load and ignores the unbalanced condition.

(B) This incorrect solution determines the roof slope factor for a slippery surface.

(D) This incorrect solution uses C_t instead of C_s calculating the sloped roof snow load.

SOLUTION 27

The shear force in the walls is a combination of the total shear force on the building and the shear induced by the torsional moment. If the resultant wind load, W, does not pass through the center of rigidity (c.r.) of the structure, the resultant creates a torsional moment that is resisted by shear in the walls.

$$V = \frac{Wr_i}{\sum r_i} + \frac{M_t x r_i}{I_p}$$

$$W = wL = \left(200 \ \frac{\text{lbf}}{\text{ft}}\right)(160 \ \text{ft})$$

$$= 32{,}000 \ \text{lbf} \quad (32 \ \text{kips})$$

$$I_p = I_{xx} + I_{yy}$$

$$I_{xx} = \sum Ay^2$$

$$I_{yy} = \sum Ax^2$$

Calculate I_{xx} using walls B and C. Calculate I_{yy} using walls A and D. x and y are the distances from the centroids of each wall area to the center of rigidity of the building.

$$I_{xx} = (20 \ \text{ft})(1 \ \text{ft})(10 \ \text{ft})^2 + (20 \ \text{ft})(1 \ \text{ft})(10 \ \text{ft})^2$$

$$= 4000 \ \text{ft}^4$$

$$I_{yy} = (50 \ \text{ft})(1 \ \text{ft})(98.3 \ \text{ft} - 65 \ \text{ft})^2$$

$$+ (40 \ \text{ft})(1 \ \text{ft})(140 \ \text{ft} - 98.3 \ \text{ft})^2$$

$$= 125{,}000 \ \text{ft}^4$$

$$I_p = 4000 \ \text{ft}^4 + 125{,}000 \ \text{ft}^4 = 129{,}000 \ \text{ft}^4$$

The torsional moment is the product of the resultant and the distance between the centroid of the load and the center of rigidity.

$$M_t = W\bar{x} = (32 \ \text{kips})(98.3 \ \text{ft} - 80 \ \text{ft}) = 586 \ \text{ft-kips}$$

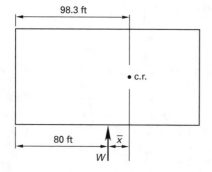

Rigidity is the relative stiffness of the walls. In this case, all walls are 12 in thick. Assume the walls all have the same modulus of elasticity and height, L_j.

$$r_j = \frac{k_j}{\sum k_i}$$

$$k_j = \frac{A_j E_j}{L_j}$$

$$r_j = \frac{k_j}{k_j + k_k} = \frac{A_j E_j}{L_j \left(\dfrac{A_j E_j}{L_j} + \dfrac{A_k E_k}{L_k}\right)}$$

Since E_j and E_k are equal and L_j and L_k are equal, the rigidity becomes

$$r_j = \frac{A_j}{A_j + A_k}$$

For wall A,

$$V_a = \frac{Wr_a}{\sum r_i} + \frac{M_t x r_a}{I_p}$$

$$= \frac{(32 \ \text{kips})(50 \ \text{ft}^2)}{50 \ \text{ft}^2 + 40 \ \text{ft}^2}$$

$$+ \frac{(586 \ \text{ft-kips})(98.3 \ \text{ft} - 65)(50 \ \text{ft}^2)}{129{,}000 \ \text{ft}^4}$$

$$= 25.3 \ \text{kips} \quad (25 \ \text{kips})$$

The answer is (D).

Why Other Options Are Wrong

(A) This incorrect solution finds the shear load on wall D instead of wall A.

(B) This incorrect solution includes the East and West walls in calculating rigidity and does not include the effects of the eccentricity of the resultant (torsional moment).

For wall A,

$$V_a = \frac{Wr_a}{\sum r_i} = \frac{(32 \ \text{kips})(50 \ \text{ft}^2)}{50 \ \text{ft}^2 + 20 \ \text{ft}^2 + 20 \ \text{ft}^2 + 40 \ \text{ft}^2}$$

$$= 12.3 \ \text{kips} \quad (12 \ \text{kips})$$

(C) This incorrect solution does not include the effects of the torsional moment.

ANALYSIS

SOLUTION 28

The size of the footing is based on service (unfactored) loads and soil pressures because footing design safety is provided by the safety factor in the allowable soil bearing pressure.

$$P = P_D + P_L$$

$$M = M_D + M_L$$

Column 1:

$$P_1 = 30 \text{ kips} + 60 \text{ kips} = 90 \text{ kips}$$
$$M_1 = 30 \text{ ft-kips} + 40 \text{ ft-kips} = 70 \text{ ft-kips}$$

Column 2:

$$P_2 = 60 \text{ kips} + 60 \text{ kips} = 120 \text{ kips}$$
$$M_2 = 30 \text{ ft-kips} + 40 \text{ ft-kips} = 70 \text{ ft-kips}$$

For the soil pressure under the footing to be uniform, the resultant load, R, must be located at the centroid of the base area.

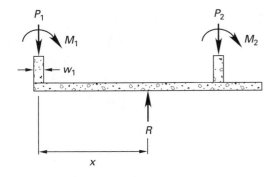

Summing moments about column 1, x is the distance to the centroid of the footing.

$$Rx = P_2(18 \text{ ft}) + M_1 + M_2$$
$$(P_1 + P_2)x = P_2(18 \text{ ft}) + M_1 + M_2$$
$$(90 \text{ kips} + 120 \text{ kips})x = (120 \text{ kips})(18 \text{ ft}) + 70 \text{ ft-kips}$$
$$+ 70 \text{ ft-kips}$$
$$(210 \text{ kips})x = 2300 \text{ ft-kips}$$
$$x = 11.0 \text{ ft}$$

The length of the footing is

$$L = \left(\tfrac{1}{2}w_1 + x\right)(2) = \left(\left(\frac{1}{2}\right)(1 \text{ ft}) + 11.0 \text{ ft}\right)(2)$$
$$= 23.0 \text{ ft}$$

The answer is (D).

Why Other Options Are Wrong

(A) This incorrect solution neglects the effects of the applied moments in calculating the location of the resultant load.

(B) This incorrect solution neglects to add one-half the column width when calculating the length of the footing.

(C) This incorrect solution uses factored loads instead of service loads to size the footing. Factored loads are used in the design of the reinforcement for the concrete footing. The size of the footing is based on service (unfactored) loads and soil pressures because footing design safety is provided by the safety factor in the allowable soil bearing pressure.

SOLUTION 29

The degree of indeterminacy of a pin-connected truss is given by the equation

$$\text{degree of indeterminacy} = 3 + \text{number of members}$$
$$- 2(\text{number of joints})$$

In this case, there is one degree of indeterminacy or redundancy.

$$\text{degree of indeterminacy} = 3 + 6 - (2)(4) = 1$$

The forces in an indeterminate truss cannot be solved directly. Since there is only one redundant member, use the dummy unit-load method to determine the force in member BC.

step 1: Draw the truss twice. Omit the redundant member on both trusses.

step 2: Load the first truss (which is now determinate) with the actual loads.

step 3: Calculate the forces, S, in all of the members. Use a positive sign for tensile forces.

step 4: Load the second truss with two unit forces acting collinearly toward each other along the line of the redundant member.

step 5: Calculate the force, u, in each of the members.

step 6: Calculate the force in the redundant member using the equation

$$S_{\text{redundant}} = \frac{-\sum\left(\dfrac{SuL}{AE}\right)}{\sum\left(\dfrac{u^2 L}{AE}\right)}$$

step 7: The true force in member j of the truss is

$$F_{j,\text{true}} = S_j + S_{\text{redundant}} u_j$$

Removing the redundant member, draw the free-body diagram of the loaded truss. The reaction loads are calculated by static analysis of the truss.

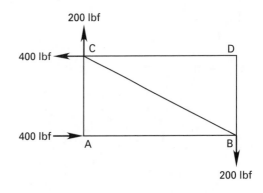

Draw the free-body diagram for each joint to determine the force in the members.

Joint A:

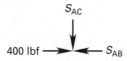

Summing the forces,

$$S_{AC} = 0 \text{ lbf}$$
$$S_{AB} = -400 \text{ lbf}$$

Joint B:

Summing the forces (see table), the horizontal component of the force in member BC must equal 400 lbf (tension). By geometry of the figure, determine that the vertical component of member BC equals 200 lbf. Therefore, the force in member BD is zero. The resultant force in member BC is

$$S_{BC} = \sqrt{(400 \text{ lbf})^2 + (200 \text{ lbf})^2}$$
$$= 447 \text{ lbf}$$

Continue in the same fashion, solving for the forces in the truss.

Draw the free-body diagram of the unit-load truss and solve for the member forces.

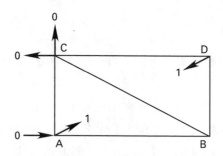

The force in redundant member AD is

$$S_{\text{redundant}} = \frac{-\sum\left(\dfrac{SuL}{AE}\right)}{\sum\left(\dfrac{u^2 L}{AE}\right)}$$

$$S_{AD} = \frac{-2191 \dfrac{\text{lbf-ft}}{\text{kip}}}{10.46 \dfrac{\text{ft}}{\text{kip}}} = -209 \text{ lbf}$$

The true force in member BC of the truss is

$$F_{\text{BC,true}} = S_{BC} + S_{\text{redundant}} u_{BC}$$
$$= 447 \text{ lbf} + (-209 \text{ lbf})(1.0)$$
$$= 238 \text{ lbf (tension)}$$

The answer is (C).

Why Other Options Are Wrong

(A) This incorrect solution calculates the force in redundant member AD instead of the force in member BC.

(B) This incorrect solution uses the same product of area and modulus of elasticity, AE, for all members of the truss.

(D) This incorrect solution calculates the force in member BC in the determinate truss due to the applied load, S_{BC}, instead of the true force in member BC, $F_{\text{BC,true}}$.

Table for Sol. 29 (solving for member forces)

member	L (ft)	AE (kips)	S (lbf)	u	$\dfrac{SuL}{AE}\left(\dfrac{\text{lbf-ft}}{\text{kip}}\right)$	$\dfrac{u^2 L}{AE}\left(\dfrac{\text{ft}}{\text{kip}}\right)$
AB	10	3	−400	−0.894	1192	2.66
AC	5	3	0	−0.447	0	0.33
CD	10	3	0	−0.894	0	2.66
BC	11.18	5	447	1.0	999	2.24
BD	5	3	0	−0.447	0	0.33
AD	11.18	5	0	1.0	0	2.24
					2191	10.46

SOLUTION 30

Moment distribution is based on the relative stiffness of the members.

The relative stiffness of each section is given as

$$K_{AB} = 0.286/\text{ft}$$

$$K_{BC} = 0.190/\text{ft}$$

The distribution factor for a joint is

$$DF = \frac{K}{\sum K}$$

$$DF_{AB} = \frac{K_{AB}}{K_{AB}} = \frac{\dfrac{0.286}{\text{ft}}}{\dfrac{0.286}{\text{ft}}}$$

$$= 1.0$$

$$DF_{BA} = \frac{K_{AB}}{K_{AB} + K_{BC}} = \frac{\dfrac{0.286}{\text{ft}}}{\dfrac{0.286}{\text{ft}} + \dfrac{0.190}{\text{ft}}}$$

$$= 0.6$$

$$DF_{BC} = \frac{K_{BC}}{K_{AB} + K_{BC}} = \frac{\dfrac{0.190}{\text{ft}}}{\dfrac{0.286}{\text{ft}} + \dfrac{0.190}{\text{ft}}}$$

$$= 0.4$$

$$DF_{CB} = 0 \quad [\text{fixed end}]$$

The fixed-end moments (FEM) as taken from a reference text are

$$FEM_{AB} = -\frac{Pb^2a}{L^2} = -\frac{(100\text{ kips})(8\text{ ft})^2(6\text{ ft})}{(14\text{ ft})^2}$$

$$= -195.9\text{ ft-kips}$$

$$FEM_{BA} = \frac{Pa^2b}{L^2} = \frac{(100\text{ kips})(6\text{ ft})^2(8\text{ ft})}{(14\text{ ft})^2}$$

$$= 146.9\text{ ft-kips}$$

$$FEM_{BC} = -\frac{wL^2}{30} = -\frac{\left(50\ \dfrac{\text{kips}}{\text{ft}}\right)(21\text{ ft})^2}{30}$$

$$= -735.0\text{ ft-kips}$$

$$FEM_{CB} = \frac{wL^2}{20} = \frac{\left(50\ \dfrac{\text{kips}}{\text{ft}}\right)(21\text{ ft})^2}{20}$$

$$= 1102.5\text{ ft-kips}$$

Use moment distribution to find the moments at the supports.

Start the moment distribution at joint B. The unbalanced moment of -588.1 ft-kips is reverse in sign and

multiplied by the distribution factors, 0.6 and 0.4. This puts a balancing correction of 352.9 ft-kips on the left side of joint B and 235.2 ft-kips on the right side of joint B, for a total balancing correction of 588.1 ft-kips. (The horizontal lines under the moments of 352.09 ft-kips and 235.2 ft-kips indicated that joint B has now been balanced.) Multiply the balancing moments by the carryover factors of 0.5 and carry the products over to the far ends of AB and BC.

Continue by moving to joint A and balancing it in a similar manner. Continue until the carryover values are small in magnitude. End when joint B is balanced.

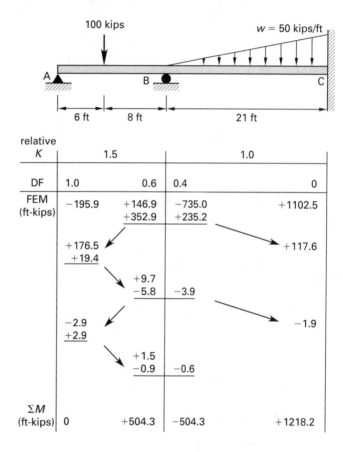

The end moments are obtained by summing all the entries at each location.

The moment at the fixed end is

$$M_C = 1102.5\text{ ft-kips} + 117.6\text{ ft-kips} - 1.9\text{ ft-kips}$$

$$= 1218.2\text{ ft-kips} \quad (1220\text{ ft-kips})$$

As a check, the moment on each side of joint B should sum to zero and the moment at the pinned end should sum to zero.

A positive moment indicates clockwise rotation at the joint, according to the sign convention used.

The answer is (B).

Why Other Options Are Wrong

(A) This incorrect solution reverses the FEMs for the triangular load. The moments on each side of joint B should be equal. In this case they do not work out.

(C) This incorrect solution solves the FEM distribution correctly as in the solution; however, the sign convention is not applied correctly and a counterclockwise rotation is assumed.

(D) This incorrect solution reverses the distribution factors at joint B, putting the distribution factor for BA at BC and vice versa.

SOLUTION 31

Find the reactions.

Member AB:

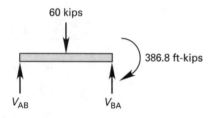

Taking clockwise moments and upward forces as positive,

$$\sum M_A = 0 \text{ ft-kips} + (60 \text{ kips})(10 \text{ ft})$$
$$+ 386.8 \text{ ft-kips} - V_{BA}(20 \text{ ft}) = 0 \text{ ft-kips}$$
$$V_{BA} = 49.3 \text{ kips}$$

Member BC:

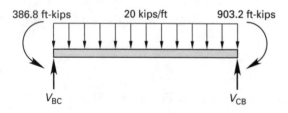

Taking clockwise moments and upward forces as positive,

$$\sum M_C = 903.2 \text{ ft-kips} - \left(20 \ \frac{\text{kips}}{\text{ft}}\right)(20 \text{ ft})(10 \text{ ft})$$
$$- 386.8 \text{ ft-kips} + V_{BC}(20 \text{ ft})$$
$$= 0$$
$$V_{BC} = 174.2 \text{ kips}$$

The reaction at B is the sum of the shears at B.

$$R_B = V_{BA} + V_{BC} = 49.3 \text{ kips} + 174.2 \text{ kips}$$
$$= 223.5 \text{ kips} \quad (220 \text{ kips})$$

The answer is (D).

Why Other Options Are Wrong

(A) This incorrect solution only considers the shear to the left of B in determining the reaction at B instead of including the shear from BC.

(B) This incorrect solution subtracts the shear forces on each side of B rather than adding them to determine the reaction at the support.

(C) This incorrect solution only considers the shear to the right of B in determining the reaction at B instead of including the shear from BA as well.

SOLUTION 32

The maximum earth pressure, p_{max}, occurs at the base of the wall.

$$p_{max} = \gamma D = \left(45 \ \frac{\text{lbf}}{\text{ft}^3}\right)(10 \text{ ft})$$
$$= 450 \text{ lbf/ft}^2$$

Draw the load, shear, and moment diagrams for the wall due to the earth pressure alone. (See the load, shear and moment diagrams.)

The roof joists bear on the full width of the concrete wall, so the roof loads are centered on the wall. The load from the storage rack can be assumed to bear at the middle of the 4 in leg of the supporting angle. The eccentricity of the rack load is

$$e = \frac{t_w}{2} + \frac{L_{bearing}}{2} = \frac{12 \text{ in}}{2} + \frac{4 \text{ in}}{2}$$
$$= 8 \text{ in}$$

The moment due to the eccentricity of the rack load is

$$M = P_{rack}e = \left(75 \ \frac{\text{lbf}}{\text{ft}}\right)(8 \text{ in})\left(\frac{1 \text{ ft}}{12 \text{ in}}\right)$$
$$= 50 \text{ ft-lbf/ft}$$

The reactions from the applied loads, including the rack load, equal 1872 lbf/ft and 377 lbf/ft at the base and top of the wall, respectively.

The shear at a distance x greater than or equal to 10 ft from the top of the wall is calculated as

$$V_x = 377 \ \frac{\text{lbf}}{\text{ft}} - \left(\left(45 \ \frac{\text{lbf}}{\text{ft}^3}\right)(x - 10 \text{ ft})\right)\frac{1}{2}(x - 10 \text{ ft})$$

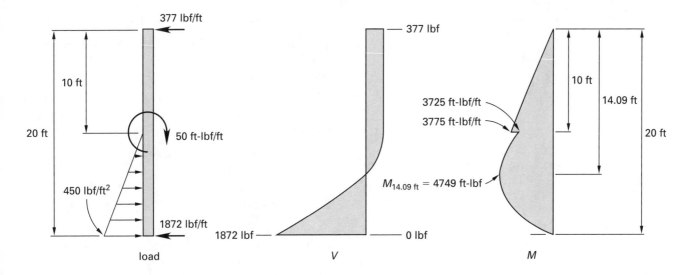

load   V   M

The maximum moment occurs at 14.09 ft from the top of the wall, where shear, V, equals 0 lbf. The moment at this point is

$$M_{\text{max}} = \left(377 \; \frac{\text{lbf}}{\text{ft}}\right)(14.09 \text{ ft})$$
$$- \left(\left(\frac{1}{2}\right)(4.09 \text{ ft})\left(45 \; \frac{\text{lbf}}{\text{ft}^3}\right)(4.09 \text{ ft})\right)$$
$$\times \left(\frac{1}{3}\right)(4.09 \text{ ft}) - 50 \text{ ft-lbf/ft}$$
$$= 4749 \text{ ft-lbf/ft} \quad (4750 \text{ ft-lbf/ft})$$

The answer is (C).

Why Other Options Are Wrong

(A) This incorrect solution mistakenly includes the roof load in the calculation of moment due to the eccentric loads.

(B) This incorrect solution does not convert the units for the eccentricity from inches to feet when calculating the moment due to the eccentric load.

(D) This incorrect solution adds the maximum moment caused by the eccentricity of the rack load to the maximum moment caused by the earth pressure.

SOLUTION 33

Section 8.3 of ACI 318 gives the moment at a given point along a one-way beam based on the moment coefficient, C_1.

$$M = C_1 w L^2$$

Moment coefficients can only be used under the following conditions: there are two or more spans; spans are approximately equal, with the larger of two adjacent spans not greater than the shorter by more than 20%; loads are uniformly distributed; the unit live load does not exceed three times the unit dead load; and members are prismatic.

The larger span (24 ft) is 20% greater than the smaller span (20 ft), so moment coefficients can be used.

The clear span, l_n, is defined in ACI 8.0 as equal to the clear span for positive moments or shear, and as the average of adjacent clear spans for negative moments.

The moment at point A (an interior face of an exterior support built integrally with the support) is

$$M_{\text{A}} = -\frac{1}{24}w_u l_n^2$$
$$= \left(-\frac{1}{24}\right)\left(600 \; \frac{\text{lbf}}{\text{ft}}\right)(24 \text{ ft})^2$$
$$= -14{,}400 \text{ ft-lbf}$$

The moment at point B (the exterior face of the first interior support) is

$$M_{\text{B}} = -\frac{1}{10}w_u l_n^2$$
$$= \left(-\frac{1}{10}\right)\left(600 \; \frac{\text{lbf}}{\text{ft}}\right)\left(\frac{24 \text{ ft} + 20 \text{ ft}}{2}\right)^2$$
$$= -29{,}040 \text{ ft-lbf}$$

The moment at point C (an interior support) is

$$M_{\text{C}} = -\frac{1}{11}w_u l_n^2$$
$$= \left(-\frac{1}{11}\right)\left(600 \; \frac{\text{lbf}}{\text{ft}}\right)\left(\frac{20 \text{ ft} + 24 \text{ ft}}{2}\right)^2$$
$$= -26{,}400 \text{ ft-lbf}$$

The moment at point D (an interior support) is

$$M_D = -\frac{1}{11}w_u l_n^2$$
$$= \left(-\frac{1}{11}\right)\left(600\ \frac{lbf}{ft}\right)(20\ ft)^2$$
$$= -21{,}818\ \text{ft-lbf}$$

The moment at point E (an exterior face of the first interior support of the other end span) is

$$M_E = -\frac{1}{10}w_u l_n^2$$
$$= \left(-\frac{1}{10}\right)\left(600\ \frac{lbf}{ft}\right)(20\ ft)^2$$
$$= -24{,}000\ \text{ft-lbf}$$

Point F is a simple support, and the moment there is zero.

The positive moment for span AB (an end span with integral support) is

$$M_{AB} = \frac{1}{14}w_u l_n^2$$
$$= \left(\frac{1}{14}\right)\left(600\ \frac{lbf}{ft}\right)(24\ ft)^2$$
$$= 24{,}686\ \text{ft-lbf}$$

The positive moment for span CD (an interior span) is

$$M_{CD} = \frac{1}{16}w_u l_n^2$$
$$= \left(\frac{1}{16}\right)\left(600\ \frac{lbf}{ft}\right)(20\ ft)^2$$
$$= 15{,}000\ \text{ft-lbf}$$

The positive moment for span EF (an end span with an unrestrained end) is

$$M_{EF} = \frac{1}{11}w_u l_n^2$$
$$= \left(\frac{1}{11}\right)\left(600\ \frac{lbf}{ft}\right)(20\ ft)^2$$
$$= 21{,}818\ \text{ft-lbf}$$

The maximum moment occurs at point B and is $-29{,}040$ ft-lbf ($-29{,}000$ ft-lbf).

The answer is (C).

Why Other Options Are Wrong

(A) This incorrect solution uses the moment coefficients for a two-span condition instead of those for a three-span condition.

(B) This incorrect solution does not use the average of adjacent clear spans when determining the negative moment at point B.

(D) This incorrect solution uses 20 ft for all three span lengths of the beam.

SOLUTION 34

Section 13.2 of ACI 318 defines a column strip as a width on each side of a column centerline equal to $0.25l_2$ or $0.25l_1$, whichever is less. A middle strip is the strip bounded by two column strips.

l_1 is defined as the length of span in the direction in which moments are being determined, measured center-to-center of the supports.

l_2 is the length of span transverse to l_1, measured center-to-center of the supports.

l_n is the length of clear span in the direction that moments are being determined, measured face-to-face of the supports.

The midspan moment of the middle strip will be a positive moment. To find the moment in the middle strip, first determine the column strip moment.

Determine the factored load on the slab. *(ACI Sec. 9.2)*

$$w_u = \overset{1.2}{\cancel{1.4}}D + \overset{1.6}{\cancel{1.7}}L$$
$$= (1.4)\left(100\ \frac{lbf}{ft^2}\right) + (1.7)\left(30\ \frac{lbf}{ft^2}\right)$$
$$= 191\ lbf/ft^2$$

The total factored static moment on the slab given in ACI 318 Sec. 13.6.2.2 is

$$M_o = \frac{w_u l_2 l_n^2}{8}$$
$$l_1 = 40\ ft$$
$$l_2 = 30\ ft$$
$$l_n = 40\ ft - (2)\left(\frac{1\ ft}{2}\right) = 39\ ft$$
$$M_o = \frac{\left(191\ \frac{lbf}{ft^2}\right)\left(\frac{1\ kip}{1000\ lbf}\right)(30\ ft)(39\ ft)^2}{8}$$
$$= \cancel{1089}\ \text{ft-kips}$$
$$\quad 958$$

The total static moment, M_o, is distributed as

$$-M_u = -0.65M_o$$
$$M_u = 0.35M_o$$

Since the midspan moment is positive, use the factored positive moment to find the positive moment in the column and middle strips.

$$M_u = 0.35M_o = (0.35)(1089 \text{ ft-kips})$$
$$= \cancel{381} \text{ ft-kips}$$
$$335$$

ACI 318 Sec. 13.6.4.4 gives the percent of positive factored moment distributed to the column strips based on the ratio of the stiffness of the beam to the stiffness of the slab, $\alpha_1 l_2/l_1$. Since this is a flat plate, there are no beams and α_1 is 0.

$$\frac{\alpha_1 l_2}{l_1} = \frac{(0)(30 \text{ ft})}{40 \text{ ft}}$$
$$= 0$$

Use the table in ACI 318 Sec. 13.6.4.4 to find the percentage of positive factored moment distributed to the column strips to be 60%.

Since the slab is symmetric, the percentage of positive moment in the middle strip is

$$M_{\text{middle},\%} = 100\% - 60\%$$
$$= 40\%$$

The positive moment in the middle strip is

$$M_{\text{middle}} = (0.40)M_u$$
$$= (0.40)(381 \text{ ft-kips})$$
$$= \cancel{152} \text{ ft-kips} \quad (\cancel{150} \text{ ft-kips})$$
$$134 \qquad\qquad 130$$

The answer is (C).

Why Other Options Are Wrong

(A) This incorrect solution uses α_1 equal to 1.0 instead of 0 when calculating the percent of positive moment distributed to the column strips.

(B) This incorrect solution reverses the values of l_1 and l_2 in calculating the total factored static moment on the slab, M_o. l_1 is in the direction in which the moments are being determined.

(D) This incorrect solution uses the center-to-center span length instead of the clear span, l_n, in determining the total factored static moment.

MECHANICS OF MATERIALS

SOLUTION 35

If the load from the joists acts within the center third of the wall cross section, the eccentricity from the load is negligible. For a 12 in wall, if the eccentricity does not exceed 2 in, the load acts within the center third.

$$e = \frac{12 \text{ in}}{2} - 2 \text{ in}$$
$$= 4 \text{ in}$$

The eccentricity of the load must be considered.

If the load acts at 12 in on center, calculate the properties of the wall.

$$A = bh = (12 \text{ in})(12 \text{ in})$$
$$= 144 \text{ in}^2$$
$$S = \frac{bh^2}{6} = \frac{(12 \text{ in})(12 \text{ in})^2}{6}$$
$$= 288 \text{ in}^3$$

The axial stress on the wall is

$$f_a = \frac{P}{A} = \frac{700 \text{ lbf}}{144 \text{ in}^2}$$
$$= 4.9 \text{ lbf/in}^2$$

The bending stress on the wall creates tension on the outside face and compression on the inside face of the wall.

$$f_b = \frac{Pe}{S} = \frac{(700 \text{ lbf})(4 \text{ in})}{288 \text{ in}^3}$$
$$= 9.7 \text{ lbf/in}^2$$

The total stress on the wall is

$$f_a + f_b = 4.9 \frac{\text{lbf}}{\text{in}^2} + 9.7 \frac{\text{lbf}}{\text{in}^2}$$
$$= 14.6 \text{ lbf/in}^2$$
$$(15 \text{ lbf/in}^2 \text{ (compression)})$$

$$f_a - f_b = 4.9 \frac{\text{lbf}}{\text{in}^2} - 9.7 \frac{\text{lbf}}{\text{in}^2}$$
$$= -4.8 \text{ lbf/in}^2$$
$$(5 \text{ lbf/in}^2 \text{ (tension)})$$

The answer is (D).

Why Other Options Are Wrong

(A) This incorrect solution neglects the effects of the eccentricity of the load (i.e., it does not calculate the bending stresses).

(B) This incorrect solution mistakenly identifies the axial stress on the wall as a tensile stress and does not consider the flexural stresses on the wall.

(C) This incorrect solution calculates the bending stresses due to the eccentricity of the load but does not add the flexural compressive stress to the axial compressive stress.

SOLUTION 36

To determine deflection by the conjugate beam method, draw the M/EI diagram for the actual load, then load the conjugate beam with the M/EI diagram. The deflection of the beam at a point is numerically equal to the moment of the conjugate beam at that point.

Draw the free-body diagram for the beam.

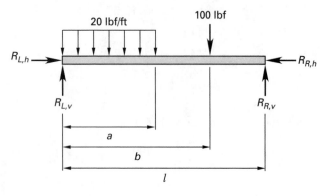

$$R_R = 95.5 \text{ lbf}$$
$$R_L = 104.5 \text{ lbf}$$

Draw the moment diagram for the beam with the actual loads.

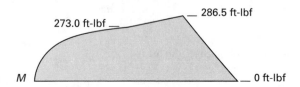

The moment at 5 ft from the left support is

$$M_{5 \text{ ft}} = R_R(6 \text{ ft}) - (100 \text{ lbf})(3 \text{ ft})$$
$$= (95.5 \text{ lbf})(6 \text{ ft}) - (100 \text{ lbf})(3 \text{ ft})$$
$$= 273.0 \text{ ft-lbf}$$

The moment at 8 ft from the left support is

$$M_{8 \text{ ft}} = R_R(3 \text{ ft})$$
$$= (95.5 \text{ lbf})(3 \text{ ft})$$
$$= 286.5 \text{ ft-lbf}$$

To determine the deflection, draw the conjugate beam loaded with the M/EI diagram.

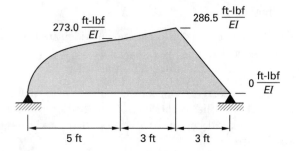

Find the equivalent loads by dividing the moment diagram into easy-to-calculate areas. The area under the curve can be approximated as the area of a parabola.

$$P_{\text{parabola}} = \tfrac{2}{3}hb$$
$$= \left(\frac{2}{3}\right)\left(\frac{273.0 \text{ ft-lbf}}{EI}\right)(5 \text{ ft})$$
$$= \frac{910 \text{ lbf-ft}^2}{EI}$$

$$P_{\text{rectangle}} = bh$$
$$= (3 \text{ ft})\left(\frac{273.0 \text{ ft-lbf}}{EI}\right)$$
$$= \frac{819 \text{ lbf-ft}^2}{EI}$$

$$P_{\text{triangle}} = \tfrac{1}{2}hb$$
$$= \left(\frac{1}{2}\right)\left(\frac{286.5 \text{ lbf-ft} - 273.0 \text{ lbf-ft}}{EI}\right)(3 \text{ ft})$$
$$= \frac{20.25 \text{ lbf-ft}^2}{EI}$$

$$P_{\text{triangle}} = \tfrac{1}{2}hb$$
$$= \left(\frac{1}{2}\right)\left(\frac{286.5 \text{ lbf-ft}}{EI}\right)(3 \text{ ft})$$
$$= \frac{429.75 \text{ lbf-ft}^2}{EI}$$

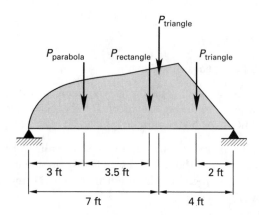

Find the conjugate reactions.

$$\sum M_R = 0 \text{ ft-kips} = \sum P_i x_i - R_L l$$
$$\sum M_R = P_{\text{triangle}}(2 \text{ ft}) + P_{\text{triangle}}(4 \text{ ft})$$
$$+ P_{\text{rectangle}}(11 \text{ ft} - 6.5 \text{ ft})$$
$$+ P_{\text{parabola}}(11 \text{ ft} - 3 \text{ ft}) - R_L(11 \text{ ft})$$

$$R_L(11 \text{ ft}) = \left(\frac{429.75 \text{ lbf-ft}^2}{EI}\right)(2 \text{ ft})$$

$$+ \left(\frac{20.25 \text{ lbf-ft}^2}{EI}\right)(4 \text{ ft})$$

$$+ \left(\frac{819.0 \text{ lbf-ft}^2}{EI}\right)(4.5 \text{ ft})$$

$$+ \left(\frac{910 \text{ lbf-ft}^2}{EI}\right)(8 \text{ ft})$$

$$= \frac{11,906 \text{ lbf-ft}^3}{EI}$$

$$R_L = \frac{1082 \text{ lbf-ft}^2}{EI}$$

The conjugate moment (deflection) at 5 ft from the left support is

$$\Delta_{5 \text{ ft}} = M_{5 \text{ ft}}$$

$$= R_L(5 \text{ ft}) - P_{\text{parabola}}(2 \text{ ft})$$

$$= \left(\frac{1082 \text{ lbf-ft}^2}{EI}\right)(5 \text{ ft}) - \left(\frac{910 \text{ lbf-ft}^2}{EI}\right)(2 \text{ ft})$$

$$= \frac{(3590 \text{ lbf-ft}^3)\left(1728 \frac{\text{in}^3}{\text{ft}^3}\right)}{\left(1.2 \times 10^6 \frac{\text{lbf}}{\text{in}^2}\right)(240 \text{ in}^4)}$$

$$= 0.0215 \text{ in} \quad (2.2 \times 10^{-2} \text{ in})$$

The answer is (C).

Why Other Options Are Wrong

(A) This incorrect solution uses the moment from the loading diagram as the conjugate moment (deflection). The units do not work out.

(B) This incorrect solution does not keep track of the units in determining the conjugate moment. The equivalent loads are calculated as lbf/EI rather than $\text{lbf-ft}^2/EI$. When calculating the resultant deflection, the unit conversion is not correct.

(D) This solution incorrectly calculates the area of the parabola in the moment diagram. The equation for the area of a full parabola is used ($P_{\text{parabola}} = (4/3)hb$) instead of that for a half-parabola.

SOLUTION 37

Section 9.5.2.3 of the ACI 318 contains the requirements for deflection calculations for beams. Immediate deflection calculations are based on service (unfactored) loads and the effective moment of inertia, I_e.

$$\Delta = \frac{5wL^4}{384E_cI_e}$$

The total uniform load is

$$w = D + L = 800 \frac{\text{lbf}}{\text{ft}} + 1200 \frac{\text{lbf}}{\text{ft}}$$

$$= 2000 \text{ lbf/ft}$$

$$E_c = 57,000\sqrt{f'_c} = (57,000)\left(\sqrt{6000} \frac{\text{lbf}}{\text{in}^2}\right)$$

$$= 4.42 \times 10^6 \text{ lbf/in}^2 \quad \text{[ACI Sec. 8.5.1]}$$

$$I_e = \left(\frac{M_{\text{cr}}}{M_a}\right)^3 I_g + \left(1 - \left(\frac{M_{\text{cr}}}{M_a}\right)^3\right) I_{\text{cr}}$$

$$\text{[ACI 318 Eq. 9-8]}$$

$$M_{\text{cr}} = \frac{f_r I_g}{y_t} \quad \text{[ACI 318 Eq. 9-9]}$$

$$f_r = 7.5\sqrt{f'_c} = (7.5)\left(\sqrt{6000} \frac{\text{lbf}}{\text{in}^2}\right)$$

$$= 581 \text{ lbf/in}^2 \quad \text{[ACI 318 Eq. 9-10]}$$

$$y_t = \frac{h}{2} = \frac{30 \text{ in}}{2}$$

$$= 15 \text{ in}$$

$$I_g = \frac{bh^3}{12} = \frac{(16 \text{ in})(30 \text{ in})^3}{12}$$

$$= 36,000 \text{ in}^4$$

$$M_{\text{cr}} = \frac{f_r I_g}{y_t}$$

$$= \frac{\left(581 \frac{\text{lbf}}{\text{in}^2}\right)(36,000 \text{ in}^4)\left(\frac{1 \text{ kip}}{1000 \text{ lbf}}\right)\left(\frac{1 \text{ ft}}{12 \text{ in}}\right)}{15 \text{ in}}$$

$$= 116 \text{ ft-kips}$$

$$M_a = \frac{wL^2}{8} = \frac{\left(2000 \frac{\text{lbf}}{\text{ft}}\right)(24 \text{ ft})^2\left(\frac{1 \text{ kip}}{1000 \text{ lbf}}\right)}{8}$$

$$= 144 \text{ ft-kips}$$

From a reference handbook, for a rectangular, singly reinforced beam,

$$I_{\text{cr}} = \frac{bc_s^3}{3} + nA_s(d - c_s)^2$$

$$c_s = n\rho d\left(\sqrt{1 + \frac{2}{n\rho}} - 1\right)$$

$$n = \frac{E_s}{E_c} = \frac{29 \times 10^6 \frac{\text{lbf}}{\text{in}^2}}{4.4 \times 10^6 \frac{\text{lbf}}{\text{in}^2}}$$

$$= 6.6$$

$$\rho = \frac{A_s}{bd} = \frac{7.90 \text{ in}^2}{(16 \text{ in})(24 \text{ in})}$$

$$= 0.0206$$

$$c_s = n\rho d \left(\sqrt{1 + \frac{2}{n\rho}} - 1 \right)$$

$$= (6.6)(0.0206)(24 \text{ in}) \left(\sqrt{1 + \frac{2}{(6.6)(0.0206)}} - 1 \right)$$

$$= 9.67 \text{ in}$$

$$I_{cr} = \frac{bc_s^3}{3} + nA_s(d - c_s)^2$$

$$= \frac{(16 \text{ in})(9.67 \text{ in})^3}{3}$$

$$+ (6.6)(7.90 \text{ in}^2)(24 \text{ in} - 9.67 \text{ in})^2$$

$$= 15{,}529 \text{ in}^4$$

$$I_e = \left(\frac{M_{cr}}{M_a} \right)^3 I_g + \left(1 - \left(\frac{M_{cr}}{M_a} \right)^3 \right) I_{cr}$$

$$= \left(\frac{116 \text{ ft-kips}}{144 \text{ ft-kips}} \right)^3 (36{,}000 \text{ in}^4)$$

$$+ \left(1 - \left(\frac{116 \text{ ft-kips}}{144 \text{ ft-kips}} \right)^3 \right) (15{,}529 \text{ in}^4)$$

$$= 26{,}230 \text{ in}^4$$

The immediate deflection is

$$\Delta = \frac{5wL^4}{384 E_c I_e}$$

$$= \frac{(5) \left(2000 \dfrac{\text{lbf}}{\text{ft}} \right) (24 \text{ ft})^4 \left(1728 \dfrac{\text{in}^3}{\text{ft}^3} \right)}{(384) \left(4.4 \times 10^6 \dfrac{\text{lbf}}{\text{in}^2} \right) (26{,}230 \text{ in}^4)}$$

$$= 0.13 \text{ in}$$

The answer is (B).

Why Other Options Are Wrong

(A) This incorrect solution uses the gross moment of inertia to calculate the deflection. This does not account for the effects of cracking.

(C) This incorrect solution uses ~~the SI equation~~ for ~~modulus of rupture instead of the U.S. equation.~~ *neglects to recognize that beam self-wt is already included in the given DL.*

(D) This incorrect solution uses factored loads in the calculations. Immediate deflection calculations should be based on unfactored loads.

SOLUTION 38

For a W12 × 65 column,

$$A = 19.1 \text{ in}^2$$
$$r_y = 3.02 \text{ in}$$
$$r_x = 5.28 \text{ in}$$
$$S_y = 29.1 \text{ in}^3$$

From AISC ASD Table C-C2.1, k is 1.0 for a column pinned at both ends. The unbraced lengths in each direction are

$$l_y = 10 \text{ ft}$$
$$l_x = 20 \text{ ft}$$

$$\left(\frac{kl}{r} \right)_y = \frac{(1.0)(10 \text{ ft}) \left(12 \dfrac{\text{in}}{\text{ft}} \right)}{3.02 \text{ in}} = 39.7$$

$$\left(\frac{kl}{r} \right)_x = \frac{(1.0)(20 \text{ ft}) \left(12 \dfrac{\text{in}}{\text{ft}} \right)}{5.28 \text{ in}} = 45.5$$

Calculate the axial stresses.

Draw the free-body diagram of the column. Summing moments about the top support,

$$\sum M = 0 \text{ ft-kips} = Pe - R_R L$$

$$= (50 \text{ kips})(8 \text{ in}) \left(\frac{1 \text{ ft}}{12 \text{ in}} \right) - R_R(20 \text{ ft})$$

$$R_R = 1.67 \text{ kips}$$
$$R_R + R_L = 0 \text{ kips}$$
$$R_L = -1.67 \text{ kips}$$

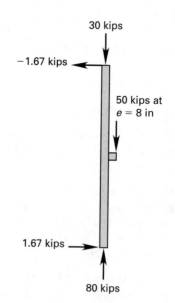

The maximum axial load, 80 kips, occurs in the lower half of the member. Allowable axial stress, F_a, is based on the largest kl/r ratio, in this case $(kl/r)_x$.

For $(kl/r)_x = 45.5$, $F_a = 18.74 \text{ kips/in}^2$ (AISC ASD Part 3, Table C-36).

The maximum axial stress is

$$f_a = \frac{P_{\text{total}}}{A} = \frac{30 \text{ kips} + 50 \text{ kips}}{19.1 \text{ in}^2}$$

$$= 4.19 \text{ kips/in}^2$$

To determine which combined stress equation applies, the ratio of actual axial stress to allowable axial stress must be determined.

$$\frac{f_a}{F_a} = \frac{4.19\ \frac{\text{kips}}{\text{in}^2}}{18.74\ \frac{\text{kips}}{\text{in}^2}} = 0.224$$

Draw the moment diagram for the column. The moment at mid-height is

$$M = R_L L = (-1.67\ \text{kips})(10\ \text{ft}) = -16.7\ \text{ft-kips}$$

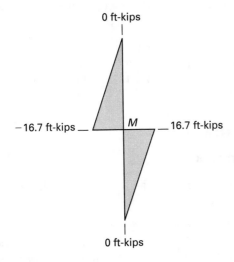

0 ft-kips

-16.7 ft-kips ____ _M_ ___ -16.7 ft-kips

0 ft-kips

The maximum moment occurs at mid-height of the column.

Calculate the bending stress.

$$f_{b,y} = \frac{M}{S_y} = \frac{(16.63\ \text{ft-kips})\left(12\ \frac{\text{in}}{\text{ft}}\right)}{29.1\ \text{in}^3}$$

$$= 6.86\ \text{kips/in}^2$$

$$f_{b,x} = 0\ \text{kips/in}^2$$

$$F_{b,x} = F_{b,y}$$

$$= 22\ \text{kips/in}^2 \quad \text{[given]}$$

$F'_{e,y}$ is based on $(kl/r)_y = 39.7$, so $F'_{e,y} = 94.79\ \text{kips/in}^2$ (AISC ASD Part 5, Table 8).

Calculate the combined stress ratios. Since f_a/F_a is greater than 0.15, use AISC ASD Eqs. H1-1 and H1-2.

$$\frac{f_a}{F_a} + \frac{C_m f_{b,y}}{\left(1 - \frac{f_a}{F'_{e,y}}\right)F_{b,y}} + \frac{C_m f_{b,x}}{\left(1 - \frac{f_a}{F'_{e,x}}\right)F_{b,x}} \le 1.0$$

[AISC ASD Eq. H1-1]

From AISC ASD Sec. H1, C_m is 1.0 for pinned members. Note that $f_{b,x}$ is zero and the third term becomes zero.

$$0.224 + \frac{(1.0)\left(6.86\ \frac{\text{kips}}{\text{in}^2}\right)}{\left(1 - \frac{4.19\ \frac{\text{kips}}{\text{in}^2}}{94.79\ \frac{\text{kips}}{\text{in}^2}}\right)\left(22\ \frac{\text{kips}}{\text{in}^2}\right)} + 0$$

$$= 0.224 + 0.326$$

$$= 0.550 \quad \text{[AISC ASD Eq. H1-1]}$$

$$\frac{f_a}{0.6F_y} + \frac{f_{b,x}}{F_{b,x}} + \frac{f_{b,y}}{F_{b,y}} \le 1.0 \quad \text{[AISC ASD Eq. H1-2]}$$

$$\frac{4.19\ \frac{\text{kips}}{\text{in}^2}}{(0.6)\left(36\ \frac{\text{kips}}{\text{in}^2}\right)} + \frac{0\ \frac{\text{kips}}{\text{in}^2}}{22\ \frac{\text{kips}}{\text{in}^2}} + \frac{6.86\ \frac{\text{kips}}{\text{in}^2}}{22\ \frac{\text{kips}}{\text{in}^2}} = 0.506$$

[AISC ASD Eq. H1-2]

The critical stress combination is 0.550, the greater of AISC ASD Eqs. H1-1 and H1-2.

The answer is (A).

Why Other Options Are Wrong

(B) This incorrect solution calculates F'_e based on $(kl/r)_x$ instead of $(kl/r)_y$. F'_e should be based on the axis about which bending occurs.

(C) In calculating the unbraced length, l_y, this incorrect solution neglects the support provided at mid-height.

(D) This incorrect solution calculates the flexural stress by including the roof load as an eccentric load and using the applied moment as the maximum moment on the column rather than by drawing the moment diagram.

Note that for the column to be adequate, the critical stress ratio must be less than or equal to 1.0.

SOLUTION 39

The torsional moment on the beam restrained at one end is

$$T = wLe = \left(20\ \frac{\text{kips}}{\text{ft}}\right)(20\ \text{ft})(9\ \text{in})\left(\frac{1\ \text{ft}}{12\ \text{in}}\right)$$

$$= 300\ \text{ft-kips}$$

The answer is (B).

Why Other Options Are Wrong

(A) This incorrect solution calculates the torsion for a beam restrained at both ends.

(C) This incorrect solution calculates the flexural moment instead of the torsional moment for a beam with both ends fully restrained (fixed).

(D) This incorrect solution calculates the flexural moment instead of the torsional moment for a simply supported beam without end restraint.

SOLUTION 40

Beams with eccentric loads are subject to torsion if the center of the load does not pass through the shear center of the shape. For symmetrical shapes (such as W shapes), the shear center is at the centroid of the shape. From the *Civil Engineering Reference Manual* or another reference, for the channel beam shown, the shear center is located at a distance from the midpoint of the vertical leg of $e = 3b^2/(h + 6b)$.

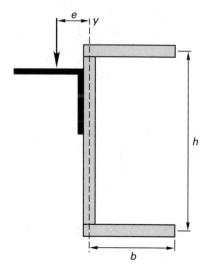

In this case,

$$b = 4.0 \text{ in} - \frac{t}{2} = 4.0 \text{ in} - \left(\frac{1}{2}\right)\left(\frac{1}{2} \text{ in}\right)$$
$$= 3.75 \text{ in}$$
$$h = 8.0 \text{ in} - t = 8.0 \text{ in} - \frac{1}{2} \text{ in}$$
$$= 7.5 \text{ in}$$
$$e = \frac{3b^2}{h + 6b} = \frac{(3)(3.75 \text{ in})^2}{7.5 \text{ in} + (6)(3.75 \text{ in})}$$
$$= 1.41 \text{ in}$$

Since the applied load acts through the shear center, there is no torsion on the beam. The load is purely flexural.

The moment due to flexure is

$$M = \frac{PL}{4} = \frac{(200 \text{ lbf})(10 \text{ ft})}{4}$$
$$= 500 \text{ ft-lbf (flexure)}$$

The answer is (D).

Why Other Options Are Wrong

(A) This incorrect solution does not check the shear center of the channel beam and assumes that the load creates a torsional moment. This solution also ignores the flexural effects of the load.

$$T = Pe = (200 \text{ lbf})(1.41 \text{ in})$$
$$= 282 \text{ in-lbf} \quad (280 \text{ in-lbf (torsion)})$$

(B) This incorrect solution does not check the shear center of the channel beam and assumes that the load creates a torsional moment. This solution also ignores the flexural effects of the load and lists the units incorrectly.

(C) This incorrect solution does not check the shear center of the channel beam and assumes that the load creates a torsional moment in addition to the flexural moment.

MATERIALS

SOLUTION 41

Since this is a bridge, the AASHTO specifications must be checked.

AASHTO *Standard Specifications for Highway Bridges*, Div. 1, Sec. 10.38.3.1 states that the effective width of the compression flange for an interior girder may not exceed the smaller of the following three measurements.

$$\frac{L}{4} = \left(\frac{30 \text{ ft}}{4}\right)\left(12 \frac{\text{in}}{\text{ft}}\right)$$
$$= 90 \text{ in}$$
$$12t = (12)(5 \text{ in})$$
$$= 60 \text{ in}$$
$$\text{beam-to-beam spacing} = (6 \text{ ft})\left(12 \frac{\text{in}}{\text{ft}}\right)$$
$$= 72 \text{ in}$$

The smallest width controls. Therefore, the effective width is equal to 60 in.

The answer is (B).

Why Other Options Are Wrong

(A) This incorrect solution limits the effective width to the flange width of the beam.

(C) This incorrect solution does not check the AASHTO requirement of $12t$, which controls in this case.

(D) This incorrect solution only considers the limitation based on the span length.

SOLUTION 42

Find the section properties of a W18 × 40 beam in Part 1 of the AISC ASD manual.

$$b_f = 6.0 \text{ in}$$
$$A = 11.8 \text{ in}^2$$
$$d = 17.90 \text{ in}$$

To find the centroid of the transformed area, determine the equivalent width of the transformed concrete.

$$b_{\text{transformed}} = \frac{1}{n} b_e = \left(\frac{1}{8}\right)(72 \text{ in})$$
$$= 9.0 \text{ in}$$

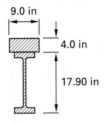

The distance to the centroid of the transformed section measured from the bottom of the beam is

$$\bar{y} = \frac{\sum A_i \bar{y}}{\sum A_i}$$

$$= \frac{(9.0 \text{ in})(4.0 \text{ in})\left(17.90 \text{ in} + \dfrac{4.0 \text{ in}}{2}\right) + (11.8 \text{ in}^2)\left(\dfrac{17.90 \text{ in}}{2}\right)}{(9.0 \text{ in})(4.0 \text{ in}) + 11.8 \text{ in}^2}$$

$$= 17.2 \text{ in} \quad (17 \text{ in})$$

The answer is (C).

Why Other Options Are Wrong

(A) This incorrect solution makes a math error in calculating the transformed area, subtracting half the thickness of the concrete instead of adding it to find the distance of the transformed concrete.

The distance to the centroid of the transformed section, measured from the bottom of the beam, is

$$\bar{y} = \frac{\sum A_i \bar{y}}{\sum A_i}$$

$$= \frac{(9.0 \text{ in})(4.0 \text{ in})\left(17.90 \text{ in} - \dfrac{4.0 \text{ in}}{2}\right) + (11.8 \text{ in}^2)\left(\dfrac{17.90 \text{ in}}{2}\right)}{(9.0 \text{ in})(4.0 \text{ in}) + 11.8 \text{ in}^2}$$

$$= 14.2 \text{ in} \quad (14 \text{ in})$$

(B) This incorrect solution uses the section properties of a W16 × 40 beam instead of a W18 × 40 beam.

(D) This incorrect solution does not transform the concrete when calculating the centroid.

SOLUTION 43

Because the concrete has not gained any of its strength immediately after pouring, the dead loads (including the slab and beam weight) are carried by the steel section alone.

The bending stress in the bottom fibers due to dead load is

$$f = \frac{M}{S}$$

$$f_{\text{bot}} = \frac{M_D}{S_x} = \frac{(80 \text{ ft-kips})\left(12 \dfrac{\text{in}}{\text{ft}}\right)}{68.4 \text{ in}^3}$$

$$= 14.0 \text{ kips/in}^2 \quad (14 \text{ kips/in}^2)$$

The answer is (B).

Why Other Options Are Wrong

(A) This incorrect solution uses the section modulus for the transformed section rather than just for the beam. The transformed section modulus would be used in shored construction.

(C) This incorrect solution uses the total moment instead of just the dead load moment.

(D) This incorrect solution uses ft-lbf for moment instead of in-lbf when calculating the stress. The units do not work out.

SOLUTION 44

Section I2, Part 2 of the AISC ASD manual limits the bending stress in an unshored steel beam with shear connectors to

$$F = 0.90 F_y = (0.90)\left(50 \dfrac{\text{kips}}{\text{in}^2}\right)$$

$$= 45.0 \text{ kips/in}^2 \quad (45 \text{ kips/in}^2)$$

The answer is (D).

Why Other Options Are Wrong

(A) This incorrect solution uses the wrong grade of steel in calculating the stress.

(B) This incorrect solution limits the bending stress to $0.66 F_y$. This limitation applies to the stress in the steel for encased beams.

(C) This incorrect solution limits the bending stress to $0.76F_y$. This limitation applies only when the beam is designed to resist, unassisted, all dead and live loads—a condition that is fairly uncommon.

SOLUTION 45

Section I4 of the AISC ASD manual states that the total horizontal shear to be resisted between the point of maximum positive moment and points of zero moment is the lesser of

$$V_h = \frac{0.85f'_c A_c}{2} = \frac{(0.85)\left(3\ \dfrac{\text{kips}}{\text{in}^2}\right)(288\ \text{in}^2)}{2}$$

$$= 367\ \text{kips} \quad [\text{AISC ASD Eq. I4-1}]$$

$$V_h = \frac{F_y A_s}{2} = \frac{\left(36\ \dfrac{\text{kips}}{\text{in}^2}\right)(11.8\ \text{in}^2)}{2}$$

$$= 212\ \text{kips} \quad [\text{AISC ASD Eq. I4-2}]$$

The horizontal shear force that must be carried by the studs is 212 kips.

The allowable horizontal shear load on a $^3/_4$ in by 3 in headed stud is found in AISC ASD Table I4.1.

$$q = 11.5\ \text{kips/stud}$$

For beams with a formed steel deck, the allowable horizontal shear load on a stud must be multiplied by the reduction factor given in AISC ASD Sec. I5.2.

$$\text{RF} = \left(\frac{0.85}{\sqrt{N_r}}\right)\left(\frac{w_r}{h_r}\right)\left(\frac{H_s}{h_r} - 1.0\right) \leq 1.0$$

From the problem statement, nominal rib height, h_r, is 2 in; length of stud connector, H_s, is 3 in; number of studs in one rib, N_r, is 1; and average width of concrete rib, w_r, is 3 in.

$$\text{RF} = \left(\frac{0.85}{\sqrt{1}}\right)\left(\frac{3\ \text{in}}{2\ \text{in}}\right)\left(\frac{3\ \text{in}}{2\ \text{in}} - 1.0\right)$$

$$= 0.64$$

The number of studs required between the point of maximum positive moment and points of zero moment is

$$n = \frac{V_h}{q\text{RF}} = \frac{212\ \text{kips}}{\left(11.5\ \dfrac{\text{kips}}{\text{stud}}\right)(0.64)}$$

$$= 28.8\ \text{studs}$$

For a simply supported beam, the maximum positive moment occurs at the midspan of the beam. Thus, the total number of shear studs required is

$$n_{\text{total}} = 2n = (2)(28.8\ \text{studs})$$

$$= 58\ \text{studs}$$

The answer is (D).

Why Other Options Are Wrong

(A) This incorrect solution does not include the reduction factor for steel decking and only calculates the number of studs required for one-half the span.

(B) This incorrect solution calculates the number of studs required for only one-half the span.

(C) This incorrect solution does not include the reduction factor for steel decking.

SOLUTION 46

Since the compressive strength of masonry is not specified, the wall must be empirically designed. Chapter 5 of the MSJC code contains the empirical requirements for thickness. Two requirements must be checked: one for lateral support and one for minimum thickness. The greater value applies.

Table 5.5.1 of the MSJC code gives wall lateral support requirements. Fully grouted bearing walls have a maximum length-to-thickness ratio (l/t) or height-to-thickness ratio (h/t) of 20. The limiting length or height is the distance between lateral supports. If the wall spans horizontally, the limiting ratio is the length-to-thickness (l/t). If the wall spans vertically, the limiting ratio is the height-to-thickness (h/t).

Since the wall spans vertically, $h/t \leq 20$.

Solving for thickness,

$$t \geq \frac{h}{20} = \frac{(10\ \text{ft})\left(12\ \dfrac{\text{in}}{\text{ft}}\right)}{20}$$

$$= 6.0\ \text{in}$$

The minimum thickness requirements of MSJC code Sec. 5.6.2 also apply. This section specifies minimum wall thickness based on the number of stories. Since the problem states that the wall spans 10 ft from the foundation to the roof, assume the wall is one story. One-story walls must have a minimum thickness of 6 in.

The answer is (A).

Why Other Options Are Wrong

(B) This incorrect solution does not recognize the wall described as a single story and uses the minimum thickness for walls more than one story (8 in).

(C) This incorrect solution uses the wall's length-to-thickness ratio as the limiting ratio. Although lateral support is provided by the intersecting walls, the problem states that the wall spans vertically. The span supports at the foundation and roof provide lateral support as well. The limiting ratio should be based on the direction of span, h/t.

(D) This incorrect solution does not recognize the problem described as being an empirically designed wall. Walls designed using allowable stress design (ASD) do not have limits on their thickness.

The fact that the compressive strength of masonry, f'_m, is not specified identifies this wall as empirically designed. Walls designed using allowable stress design must specify f'_m.

SOLUTION 47

Since the support condition of the beam is not known, the extreme fiber stress in tension in the concrete is limited, in Sec. 18.4.1 of ACI 318, to

$$\sigma = 3\sqrt{f'_{ci}} = (3)\left(\sqrt{4000} \ \frac{\text{lbf}}{\text{in}^2}\right)$$
$$= 189.7 \ \text{lbf/in}^2 \quad (190 \ \text{lbf/in}^2)$$

The answer is (B).

Why Other Options Are Wrong

(A) This incorrect solution identifies the initial prestress force in the tendon, P_i, as the tensile stress. Note that the units for initial prestress force are actually kips, not lbf/in^2.

(C) This incorrect solution uses specified strength of concrete rather than the concrete strength at the time of initial prestress when calculating the allowable stress.

(D) This incorrect solution uses the stress limit for simply supported members found in ACI 318 Sec. 18.4.1(c).

SOLUTION 48

The nominal shear strength provided by the concrete alone is

$$V_c = 2\sqrt{f'_c}b_w d = (2)\left(\sqrt{3000} \ \frac{\text{lbf}}{\text{in}^2}\right)(6 \ \text{in})(7.5 \ \text{in})$$
$$ \overset{3700}{= 4929.5 \ \text{lbf}}$$
$$\phi V_c = \overset{0.75}{0.85}V_c = (0.85)(4929.5 \ \text{lbf})$$
$$= 4190 \ \text{lbf}$$

ACI 318 Sec. 11.5.5.1 states that if $V_u < \phi V_c/2$ and $h < 10 \ \text{in}$, then no shear reinforcement is required.

$$V_u = 2000 \ \text{lbf} < \frac{\phi V_c}{2} = \frac{4190 \ \text{lbf}}{2}$$
$$\overset{1800}{= 2095 \ \text{lbf}} \quad < 1850 \ \text{lbf}$$
$$h = 9 \ \text{in} < 10 \ \text{in}$$

Since both conditions are met, no shear reinforcement is required.

The answer is (D).

Why Other Options Are Wrong

(A) This incorrect solution misses the exception, given in ACI 318 Sec. 11.5.5.1, to the minimum shear requirement and mistakenly uses the minimum shear reinforcement requirement found in Sec. 11.5.5.3.

(B) This incorrect solution misses the exception, given in ACI 318 Sec. 11.5.5.1, to the minimum shear requirement and uses h for d when calculating the maximum stirrup spacing.

(C) This incorrect solution misses the exception, given in ACI 318 Sec. 11.5.5.1, to the minimum shear requirement and incorrectly uses the largest stirrup spacing found in ACI 318 Sec. 11.5.4 instead of the smallest.

SOLUTION 49

ACI 318 Sec. 12.2.2 contains the development length requirements for bars in tension. The development length is measured from the point where the stress in the bars is at maximum. For positive moment, the distance is measured from the center of the span. Since the clear spacing of the bars is greater than twice the diameter of the bars and the clear cover is greater than the diameter of the bars, the development length for a no. 9 bar is given as

$$l_d = \left(\frac{f_y \alpha \beta \lambda}{20\sqrt{f'_c}}\right)d_b$$
$$d_b = 1.128 \ \text{in}$$

ACI 318 Sec. 12.2.4 gives the factors used in this equation.

$$\alpha = 1.0$$
$$\beta = 1.0$$
$$\lambda = 1.0$$
$$l_d = \left(\frac{60{,}000 \ \frac{\text{lbf}}{\text{in}^2}}{(20)\left(\sqrt{6000} \ \frac{\text{lbf}}{\text{in}^2}\right)}\right)(1.128 \ \text{in})$$
$$= 43.7 \ \text{in}$$

ACI 318 Sec. 12.10.3 states that reinforcement shall extend beyond the point at which it is no longer needed for a distance equal to the greater of one of the following.

$$d = h - \text{cover} - \frac{d_b}{2}$$
$$= 30 \ \text{in} - 1.5 \ \text{in} - \frac{1.128 \ \text{in}}{2}$$
$$= 27.9 \ \text{in}$$
$$12d_b = (12)(1.128 \ \text{in})$$
$$= 13.5 \ \text{in}$$

The point at which the positive moment reinforcement is no longer needed is the point of inflection. Since the point of inflection is 3 ft from the face of the support, the bars must extend to within 36 in minus 27.9 in of the face of the support, which equals 8.1 in.

ACI 318 Sec. 12.11.1 states that at least $\frac{1}{4}$ of the positive moment reinforcement must extend into the support at least 6 in. This requirement controls.

The answer is (B).

Why Other Options Are Wrong

(A) This incorrect solution assumes that since there is no positive moment at the support, no reinforcement is required. However, ACI 318 Sec. 12.11.1 says otherwise.

(C) This incorrect solution assumes that the point where the positive moment reinforcement is no longer needed is the face of the support. The point of inflection is actually the point where the positive moment goes to zero.

(D) This incorrect solution assumes the development length to be measured from the face of the support. Development length should be measured from the point where the stress in the bars is at maximum. For positive moment, the distance should be measured from the center of the span.

SOLUTION 50

The first part of the designation, 20F, indicates the tensile capacity in bending, which is 2000 lbf/in².

The meaning of the combination symbols for glulam timbers can be found in AITC 117 and in textbooks on wood and timber design.

The answer is (D).

Why Other Options Are Wrong

(A) This answer is incorrect. The "V" in the combination symbol indicates that the glulam beam is visually graded. Mechanically graded is indicated with an "E" designation.

(B) This answer is incorrect. The depth of the beam is as designed and is not indicated in the combination symbol.

(C) This answer incorrectly assumes the shear capacity is indicated by V3. The "V" indicates that the member is visually graded. The "3" is part of the combination symbol and is not an indication of strength.

SOLUTION 51 (Use f_b instead of F_b)

Begin by determining the section properties of the glulam rafters. Table 1D in the NDS Supplement lists the section properties of the $8\frac{1}{2}$ in by $27\frac{1}{2}$ in southern pine glulam timber.

$$A = 233.8 \text{ in}^2$$
$$S_x = 1071 \text{ in}^3$$

The allowable bending design value, F_b', is given in NDS Table 2.3.1 as

$$F_b' = F_{b,xx}C_DC_MC_tC_LC_{\text{fu}}C_cC_f$$

or

$$F_b' = F_{b,xx}C_DC_MC_tC_VC_{\text{fu}}C_cC_f$$

The smaller of the two equations is the critical one because C_V and C_L shall not apply simultaneously for glulam timber bending members, according to NDS Sec. 5.3.6.

Not all of the adjustment factors are applicable in this case. The wet service factor, C_M, and the temperature factor, C_t, do not apply because the environment is neither wet nor hot. The flat-use factor, C_{fu}, does not apply because the rafters are not used in this way (NDS Sec. 5.3.7). The curvature factor, C_c, does not apply because the rafters are not curved. The form factor, C_f, only applies to members with a circular or square cross section loaded on the diagonal.

The allowable bending design value, F_b', with the applicable adjustment factors is

$$F_b' = F_{b,xx}C_DC_L$$

or

$$F_b' = F_{b,xx}C_DC_V$$

Calculate the beam stability factor, C_L, and the volume factor, C_V. The smaller value will be used to calculate the allowable bending design value.

The beam stability factor is 1.0 when the compression edge of a bending member is supported throughout its length to prevent lateral displacements and when the ends at points of bearing have lateral support to prevent rotation (NDS Sec. 3.3.3.3).

$$C_V = \left(\frac{21}{L}\right)^{1/x}\left(\frac{12}{d}\right)^{1/x}\left(\frac{5.125}{b}\right)^{1/x} \leq 1.0$$

[NDS Eq. 5.3-1]

L is the length of the bending member between points of zero moment, in feet.

x is 20 for southern pine and 10 for all other species.

In this case, L is 15 ft, d is 27.5 in, b is 8.5 in, and x is 20.

$$C_V = \left(\frac{21}{L}\right)^{1/x} \left(\frac{12}{d}\right)^{1/x} \left(\frac{5.125}{b}\right)^{1/x}$$

$$= \left(\frac{21}{15}\right)^{1/20} \left(\frac{12}{27.5}\right)^{1/20} \left(\frac{5.125}{8.5}\right)^{1/20}$$

$$= 0.95$$

Since C_V is less than C_L, use the equation

$$F_b' = F_{b,xx} C_D C_V$$

Two load cases must be examined to determine the critical stress ratio—dead load plus live load $(D + L)$, and dead load plus live load plus wind load $(D + L + W)$.

Dead Load plus Live Load:

C_D is 1.0 for $D + L$, according to App. B of NDS Sec. B.2.

$$F_b = \frac{M}{S}$$

$$F_{b,D+L} = \frac{M_D + M_L}{S_x} = \frac{60{,}000 \text{ ft-lbf} + 90{,}000 \text{ ft-lbf}}{1071 \text{ in}^3}$$

$$= \left(\frac{150{,}000 \text{ ft-lbf}}{1071 \text{ in}^3}\right)\left(12 \frac{\text{in}}{\text{ft}}\right)$$

$$= 1681 \text{ lbf/in}^2$$

$$F_b' = F_{b,xx} C_D C_V = \left(2000 \frac{\text{lbf}}{\text{in}^2}\right)(1.0)(0.95)$$

$$= 1900 \text{ lbf/in}^2$$

$$\frac{F_{b,D+L}}{F_b'} = \frac{1681 \dfrac{\text{lbf}}{\text{in}^2}}{1900 \dfrac{\text{lbf}}{\text{in}^2}}$$

$$= 0.88$$

Dead Load plus Live Load plus Wind Load:

C_D is 1.6 for $D + L + W$, according to App. B of NDS Sec. B.2.

$$F_b = \frac{M}{S}$$

$$F_{b,D+L+W} = \frac{M_D + M_L + M_W}{S_x}$$

$$= \left(\frac{\begin{array}{c}60{,}000 \text{ ft-lbf} + 90{,}000 \text{ ft-lbf}\\ + 100{,}000 \text{ ft-lbf}\end{array}}{1071 \text{ in}^3}\right)$$

$$\times \left(12 \frac{\text{in}}{\text{ft}}\right)$$

$$= 2801 \text{ lbf/in}^2$$

Calculate the allowable bending design value. The NDS allows the use of both the load duration factor, C_D, and the one-third stress increase for wind loading.

$$F_b' = 1.33 F_{b,xx} C_D C_V$$

$$= (1.33)\left(2000 \frac{\text{lbf}}{\text{in}^2}\right)(1.6)(0.95)$$

$$= 4043 \text{ lbf/in}^2$$

$$\frac{F_{b,D+L+W}}{F_b'} = \frac{2801 \dfrac{\text{lbf}}{\text{in}^2}}{4043 \dfrac{\text{lbf}}{\text{in}^2}} = 0.69$$

The critical ratio is the larger of the two, 0.88.

The answer is (B).

Why Other Options Are Wrong

(A) This incorrect solution neglects to include the volume adjustment factor, C_V, in calculating the allowable bending design value.

(C) This incorrect solution does not include the one-third allowable stress increase for wind loading allowed by the code when calculating the allowable bending design value, F_b'.

(D) This incorrect solution does not multiply the tabulated bending value by the necessary adjustment factors to determine the allowable bending design value, F_b'.

MEMBER DESIGN

SOLUTION 52

The span length of the lintel is

$$L = \text{clear span} + \frac{1}{2}\sum \text{bearing length at each end}$$

$$= 6.0 \text{ ft} + \left(\frac{1}{2}\right)(8 \text{ in} + 8 \text{ in})\left(\frac{1 \text{ ft}}{12 \text{ in}}\right)$$

$$= 6.67 \text{ ft}$$

Calculate the loads on the lintel.

The total uniform load is the sum of the dead and live loads plus the lintel self-weight.

$$w_{\text{uniform}} = w_D + w_L + w_{\text{self-wt}}$$
$$= 450 \; \frac{\text{lbf}}{\text{ft}} + 550 \; \frac{\text{lbf}}{\text{ft}} + 120 \; \frac{\text{lbf}}{\text{ft}}$$
$$= 1120 \; \text{lbf/ft}$$

The total triangular load is

$$W = \tfrac{1}{2} Lh$$
$$= \left(\frac{1}{2} \right) (6.67 \text{ ft}) \left(300 \; \frac{\text{lbf}}{\text{ft}} \right)$$
$$= 1000 \text{ lbf}$$

Calculate the design moment.

$$M_{\text{total}} = M_{\text{uniform}} + M_{\text{triangular}}$$
$$= \frac{wL^2}{8} + \frac{WL}{6}$$
$$= \left(\frac{\left(1120 \; \frac{\text{lbf}}{\text{ft}} \right) (6.67 \text{ ft})^2}{8} \right) \left(12 \; \frac{\text{in}}{\text{ft}} \right)$$
$$\quad + \left(\frac{(1000 \text{ lbf})(6.67 \text{ ft})}{6} \right) \left(12 \; \frac{\text{in}}{\text{ft}} \right)$$
$$= 88{,}080 \text{ in-lbf}$$

Determine the effective depth to reinforcement, d. The two bars side-by-side at the bottom of the beam are shown.

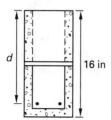

Allowing for the thickness of the unit (1.25 in) and one-half the bar diameter (0.25 in) plus grout cover (0.5 in), estimate d.

$$d = 16 \text{ in} - (1.25 \text{ in} + 0.25 \text{ in} + 0.5 \text{ in})$$
$$= 14 \text{ in}$$

Determine the stress in the steel using the equations found in a reference handbook on masonry walls.

$$f_s = \frac{M}{A_s j d}$$
$$\rho = \frac{A_s}{bd}$$
$$k = \sqrt{2\rho n + (\rho n)^2} - \rho n$$
$$j = 1 - \frac{k}{3}$$
$$n = \frac{E_s}{E_m}$$

MSJC code Sec. 1.8 specifies the moduli of elasticity of steel and masonry, respectively, as

$$E_s = 29 \times 10^6 \; \text{lbf/in}^2$$
$$E_m = 900 f'_m \quad \text{[for concrete masonry]}$$
$$= (900) \left(3000 \; \frac{\text{lbf}}{\text{in}^2} \right)$$
$$= 2.7 \times 10^6 \; \text{lbf/in}^2$$
$$n = \frac{E_s}{E_m} = \frac{29 \times 10^6 \; \frac{\text{lbf}}{\text{in}^2}}{2.7 \times 10^6 \; \frac{\text{lbf}}{\text{in}^2}} = 10.7$$

The allowable tensile stress for Grade 60 reinforcement is given in MSJC code Sec. 2.3.

$$F_s = 24{,}000 \; \text{lbf/in}^2$$

Given two no. 4 bars, App. F of ACI 318 lists the area of a no. 4 reinforcing bar as 0.20 in^2.

$$A_{s_{\text{prov}}} = (2)(0.20 \text{ in}^2)$$
$$= 0.40 \text{ in}^2$$
$$\rho = \frac{A_{s_{\text{prov}}}}{bd} = \frac{0.40 \text{ in}^2}{(7.5 \text{ in})(14 \text{ in})}$$
$$= 0.00381$$
$$k = \sqrt{2\rho n + (\rho n)^2} - \rho n$$
$$= \sqrt{(2)(0.00381)(10.7) + \big((0.00381)(10.7) \big)^2}$$
$$\quad - (0.00381)(10.7)$$
$$= 0.248$$
$$j = 1 - \frac{k}{3} = 1 - \frac{0.248}{3}$$
$$= 0.917$$

The stress in the steel is

$$f_s = \frac{M}{A_s j d} = \frac{88{,}080 \text{ in-lbf}}{(0.40 \text{ in}^2)(0.917)(14 \text{ in})}$$
$$= 17{,}152 \; \text{lbf/in}^2 \quad (17{,}200 \; \text{lbf/in}^2)$$

The allowable tensile stress in the steel is

$$F_s = 24{,}000 \; \text{lbf/in}^2 > f_s$$
$$= 17{,}200 \; \text{lbf/in}^2 \quad \text{[OK]}$$

The answer is (D).

Why Other Options Are Wrong

(A) This incorrect solution uses the clear span for the span length. Span length should include one-half the length of bearing on each side of the opening.

(B) This incorrect solution uses the depth of the lintel (16 in) for d instead of the effective depth to the reinforcement.

(C) This incorrect solution disregards the self-weight of the lintel in calculating the loads.

SOLUTION 53

Section 11.6 of ACI 318 covers the design of beams with torsion. First, check to see if torsion must be considered. ACI 318 Sec. 11.6.1 specifies that torsion can be neglected if

$$T_u < \frac{\phi T_{cr}}{4}$$

$$T_{cr} = 4\sqrt{f_c'}\left(\frac{A_{cp}^2}{p_{cp}}\right)$$

The area enclosed by the outside perimeter of concrete is given as

$$A_{cp} = 1400 \text{ in}^2$$

The outside perimeter of the concrete cross section is given as

$$p_{cp} = 150 \text{ in}$$

$$T_{cr} = 4\sqrt{f_c'}\left(\frac{A_{cp}^2}{p_{cp}}\right)$$

$$= (4)\left(\sqrt{4000}\,\frac{\text{lbf}}{\text{in}^2}\right)\left(\frac{(1400 \text{ in}^2)^2}{150 \text{ in}}\right)$$

$$\times \left(\frac{1 \text{ kip}}{1000 \text{ lbf}}\right)\left(\frac{1 \text{ ft}}{12 \text{ in}}\right)$$

$$= 275.5 \text{ ft-kips}$$

$$\phi = 0.85 \quad \overset{0.75}{} \quad [\text{ACI 318 Sec. 9.3.2.3}]$$

$$\frac{\phi T_{cr}}{4} = \frac{(0.85)(275.5 \text{ ft-kips})}{4}$$

$$\overset{51.7}{= 58.5} \text{ ft-kips}$$

$$T_u = 400 \text{ ft-kips} > \frac{\phi T_{cr}}{4} \quad [\text{must consider torsion}]$$

The total area of stirrups required is the sum of the area required for torsion and the area required for shear.

The transverse reinforcement required for torsion is given by ACI 318 Eq. 11-21.

$$\frac{A_t}{s} = \frac{T_n}{2A_o f_{y,v}\cot\theta} = \frac{\dfrac{T_u}{\phi}}{2A_o f_{y,v}\cot\theta}$$

The area enclosed by the centerline of the torsional reinforcement is

$$A_{oh} = b_t d_t = (31 \text{ in})(36 \text{ in})$$

$$= 1116 \text{ in}^2$$

$$A_o = 0.85 A_{oh} = (0.85)(1116 \text{ in}^2)$$

$$= 948.6 \text{ in}^2$$

$$\theta = 45° \quad [\text{ACI 318 Sec. 11.6.3.6}]$$

$$\frac{A_t}{s} = \frac{\left(\dfrac{400 \text{ ft-kips}}{0.85\,0.75}\right)\left(12\,\dfrac{\text{in}}{\text{ft}}\right)\left(1000\,\dfrac{\text{lbf}}{\text{kip}}\right)}{(2)(948.6 \text{ in}^2)\left(40{,}000\,\dfrac{\text{lbf}}{\text{in}^2}\right)\cot 45}$$

$$= 0.0744 \text{ in}^2/\text{in per leg}$$

The reinforcement required for shear is given by ACI 318 Eq. 11-2.

$$V_s = V_n - V_c = \frac{V_u}{\phi} - V_c$$

$$= \frac{\overset{163}{185} \text{ kips}}{\underset{0.75}{0.85}} - 168.2 \text{ kips}$$

$$\overset{49.1}{= 49.4} \text{ kips}$$

Using ACI 318 Eq. 11-15,

$$\frac{A_v}{s} = \frac{V_s}{f_y d} = \frac{(49.4 \text{ kips})\left(1000\,\dfrac{\text{lbf}}{\text{kip}}\right)}{\left(40{,}000\,\dfrac{\text{lbf}}{\text{in}^2}\right)(38 \text{ in})}$$

$$\overset{0.0323}{= 0.0325} \text{ in}^2/\text{in}$$

Because the area of shear reinforcement is divided between two vertical legs, the total area of stirrups required is

$$A_{req} = \frac{A_t}{s} + \frac{A_v}{2s} = 0.0744\,\frac{\text{in}^2}{\text{in}} + \frac{0.0325\,\dfrac{\text{in}^2}{\text{in}}}{2}$$

$$= 0.0906 \text{ in}^2/\text{in}$$

The minimum area of transverse closed stirrups is given by ACI 318 Eq. 11-23,

$$A_v + 2A_t = 0.75\sqrt{f_c'}\,\frac{b_w s}{f_{y,v}}$$

Rearranging and dividing both sides by $2s$,

$$\frac{1}{2}\left(\frac{2A_t}{s} + \frac{A_v}{s}\right) = \frac{1}{2}\left(\frac{0.75\sqrt{f_c'}b_w}{f_{y,v}}\right)$$

$$A_{min} = \frac{A_t}{s} + \frac{A_v}{2s}$$

$$= \left(\frac{1}{2}\right)\left(\frac{(0.75)\left(\sqrt{4000}\,\dfrac{\text{lbf}}{\text{in}^2}\right)(35 \text{ in})}{40{,}000\,\dfrac{\text{lbf}}{\text{in}^2}}\right)$$

$$= 0.0207 \text{ in}^2/\text{in} < 0.0906 \text{ in}^2/\text{in} \quad [\text{OK}]$$

Using no. 5 stirrups, A is 0.31 in^2 per leg.

The spacing required is

$$s = \frac{A}{A_{req}} = \frac{0.31 \text{ in}^2}{0.0906\,\dfrac{\text{in}^2}{\text{in}}}$$

$$= 3.42 \text{ in}$$

The maximum spacing permitted by ACI 318 Sec. 11.6.6.1 for torsion is the lesser of $p_h/8$ or 12 in.

p_h is the perimeter of the area enclosed by the centerline of the transverse reinforcement.

$$p_h = 2b_t + 2d_t = (2)(31 \text{ in}) + (2)(36 \text{ in})$$
$$= 134 \text{ in}$$
$$\frac{p_h}{8} = \frac{134 \text{ in}}{8} = 16.75 \text{ in}$$

12 in spacing controls for torsion. The maximum spacing permitted by ACI 318 Sec. 11.5.4.1 for shear is the lesser of $d/2$ or 24 in.

$$\frac{d}{2} = \frac{38 \text{ in}}{2} = 19 \text{ in}$$

The 12 in spacing limit controls for both shear and torsion and is greater than the required spacing. Therefore, s is ~~3.42~~ in (~~3.4~~ in).

3.09   3.1

The answer is (C). ✓

Why Other Options Are Wrong

(A) This incorrect solution uses the area of a no. 4 stirrup instead of a no. 5 in calculating the required spacing.

(B) This incorrect solution applies the minimum reinforcement requirement to the shear reinforcement alone. The minimum area of reinforcement should be checked against the combined shear and torsional reinforcement. (In this case, this incorrect value can also be calculated if the maximum torsion and the maximum shear values for the entire span are combined.)

(D) This incorrect solution sizes the stirrups for torsion only and does not check the minimum reinforcement requirement found in ACI 318 Sec. 11.6.5.2.

SOLUTION 54

The longitudinal reinforcement for beams with flexure and torsion is the sum of the area of reinforcement required for flexure and the area required for torsion.

ACI 318 Sec. 11.6.3.7 gives the additional longitudinal reinforcement required for torsion as

$$A_l = \left(\frac{A_t}{s}\right) p_h \left(\frac{f_{y,v}}{f_{y,l}}\right) \cot^2 \theta$$

p_h, the perimeter of the area enclosed by the centerline of the transverse reinforcement, is given as

$$p_h = 134 \text{ in}$$

From ACI 318 Sec. 11.6.3.6, θ is 45°.

$$A_l = \left(0.0744 \frac{\text{in}^2}{\text{in}}\right)(134 \text{ in})\left(\frac{40{,}000 \frac{\text{lbf}}{\text{in}^2}}{60{,}000 \frac{\text{lbf}}{\text{in}^2}}\right)\cot^2 45$$
$$= 6.65 \text{ in}^2$$

ACI 318 Sec. 11.6.5.3 gives the minimum area of longitudinal torsional reinforcement as

$$A_{l,\min} = \frac{5\sqrt{f'_c}A_{cp}}{f_{y,l}} - \left(\frac{A_t}{s}\right) p_h \left(\frac{f_{y,v}}{f_{y,l}}\right)$$

[ACI 318 Eq. 11-24]

$$A_{cp} = bh = (35 \text{ in})(40 \text{ in})$$
$$= 1400 \text{ in}^2$$

$$A_{l,\min} = \frac{(5)\left(\sqrt{4000 \frac{\text{lbf}}{\text{in}^2}}\right)(1400 \text{ in}^2)}{60{,}000 \frac{\text{lbf}}{\text{in}^2}}$$

$$- \left(0.0744 \frac{\text{in}^2}{\text{in}}\right)(134 \text{ in})\left(\frac{40{,}000 \frac{\text{lbf}}{\text{in}^2}}{60{,}000 \frac{\text{lbf}}{\text{in}^2}}\right)$$

$$= 0.73 \text{ in}^2 < A_l \quad [\text{OK}]$$

The area of longitudinal torsional reinforcement (6.65 in²) must be distributed equally around the perimeter of the beam at a maximum spacing of 12 in (ACI 318 Sec. 11.6.6.2). Therefore, a minimum of 12 bars is needed: four at the top, four at the bottom, and two on each side.

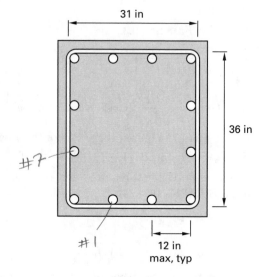

The area of reinforcement required per bar can be calculated as

$$A_l = \frac{6.65 \text{ in}^2}{12 \text{ bars}} = 0.55 \text{ in}^2/\text{bar}$$

This amount must be added to the area of reinforcement required for flexure.

Several methods can be used to determine the required flexural steel. Using a design aid found in *Design of Concrete Structures* by Nilson and Winter or another reference, calculate

$$\frac{M_u}{\phi b d^2} = \frac{(500 \text{ ft-kips})\left(12{,}000 \frac{\text{in-lbf}}{\text{ft-kip}}\right)}{(0.9)(35 \text{ in})(38 \text{ in})^2} = 132 \text{ lbf/in}^2$$

This value corresponds to an area of reinforcement that is below the minimum area required.

$$A_{s,\min} = \rho_{\min} b d = \left(\frac{200}{f_y}\right) b d$$

$$= \left(\frac{200}{60{,}000 \frac{\text{lbf}}{\text{in}^2}}\right)(35 \text{ in})(38 \text{ in})$$

$$= 4.43 \text{ in}^2$$

The total longitudinal reinforcement in the bottom of the beam is the area required for flexure plus the area of the four bars required for torsion.

$$A_s = A_l + A_{s,\min}$$

$$= (4 \text{ bars})\left(0.55 \frac{\text{in}^2}{\text{bar}}\right) + 4.43 \text{ in}^2$$

$$= 6.63 \text{ in}^2$$

The area of reinforcement per bar is

$$A_{s/\text{bar}} = \frac{A_s}{\text{number of bars}} = \frac{6.63 \text{ in}^2}{4 \text{ bars}}$$

$$= 1.66 \text{ in}^2/\text{bar}$$

A no. 11 bar provides 1.56 in² of steel. A no. 14 bar provides 2.25 in² of steel. However, no. 14 bars are not normally used for beams. Determine the size of the bars if five bars are used along the top and bottom of the beam and two bars are used on each side.

$$A_l = \frac{6.65 \text{ in}^2}{14 \text{ bars}} = 0.48 \text{ in}^2/\text{bar}$$

The total longitudinal reinforcement in the bottom of the beam is the area required for flexure plus the area of the five bars required for torsion.

$$A_s = A_l + A_{s,\min}$$

$$= (5 \text{ bars})\left(0.48 \frac{\text{in}^2}{\text{bar}}\right) + 4.43 \text{ in}^2$$

$$= 6.83 \text{ in}^2$$

The area of reinforcement per bar is

$$A_{s/\text{bar}} = \frac{A_s}{\text{number of bars}} = \frac{6.83 \text{ in}^2}{5 \text{ bars}}$$

$$= 1.37 \text{ in}^2/\text{bar}$$

A no. 11 bar provides 1.56 in² of steel. Use five no. 11 bars on the bottom of the beam. Provide uniformly distributed bars around the perimeter of the cross section, A_l, using no. 7 bars.

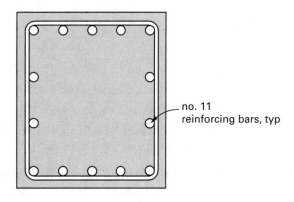

no. 11 reinforcing bars, typ

The answer is (C).

Why Other Options Are Wrong

(A) This incorrect solution calculates the reinforcement required for flexure only.

(B) This incorrect solution uses 36 in for d in calculating the amount of flexural reinforcement required.

(D) This incorrect solution uses no. 14 bars. The largest bar size usually used in beams is a no. 11 bar. Number 14 bars and larger are typically reserved for use in columns.

SOLUTION 55

ACI 318 Sec. 13.6.5 covers the moments in beams in two-way slab systems.

$$\frac{\alpha_1 l_2}{l_1} = \frac{(1.5)(22 \text{ ft})}{24 \text{ ft}}$$

$$= 1.38$$

Since 1.38 is greater than 1.0, the beam must be proportioned to resist 85% of the column strip moments plus the weight of the beam stem.

$$M_{u,\text{bm}} = 0.85 M_{u,c \text{ strip}} + M_{u,\text{bm weight}}$$

The midspan moment of a uniformly loaded two-span beam is

$$M_{u,\text{bm weight}} = \frac{w_{u,\text{bm weight}} L^2}{14}$$

$$w_{u,\text{bm weight}} = 1.4 \gamma b h$$

$$= (1.4)\left(150 \frac{\text{lbf}}{\text{ft}^3}\right)(1 \text{ ft})(2 \text{ ft})$$

$$= 420 \text{ lbf/ft}$$

$$M_{u,\mathrm{bm}} = 0.85 M_{u,c \; \mathrm{strip}} + M_{u,\mathrm{bm \; weight}}$$
$$= (0.85)(90 \text{ ft-kips})$$
$$+ \frac{\left(420 \; \frac{\text{lbf}}{\text{ft}}\right)\left(\frac{1 \text{ kip}}{1000 \text{ lbf}}\right)(24 \text{ ft})^2}{14}$$
$$= 94 \text{ ft-kips}$$

The answer is (C).

Why Other Options Are Wrong

(A) This incorrect solution does not add the self-weight of the beam to the total moment on the beam.

(B) This incorrect solution does not apply the load factor to the beam weight when calculating the moment on the beam.

(D) This incorrect solution calculates the moment on the beam using a tributary area equal to one-half the distance to each column instead of proportioning the beam moment based on the column strip moment as specified in ACI 318 Sec. 13.6.5. Additionally, it does not use the factored load.

SOLUTION 56

The allowable shear stress on a fillet weld is given in Table J2.5 of the AISC ASD manual as

$$F_v = 0.30(\text{nominal tensile strength of the weld metal})$$
$$= 0.30 F_{u,\mathrm{rod}}$$

The nominal tensile strength of the weld metal is determined by the electrodes used. E70XX electrodes have a nominal tensile strength of 70 kips/in^2.

→ *The length of the weld = nominal dimension of tube less the radius @ each end.*

The shear capacity of a weld is

The outside corner radius = 2 × wall thickness.

$$R_w = 0.30 t_e F_{u,\mathrm{rod}} L_w$$

$L_w = 2(4'' - 2 \times 0.375'')$
$= 5''$

$$t_e = 0.707 w \quad [\text{for fillet welds using SMAW}]$$

Use $L_w = 5''$
$t = 0.16''$

$$R_w = 0.30 t_e F_{u,\mathrm{rod}} L_w$$
$$= 0.30 t_e \left(70 \; \frac{\text{kips}}{\text{in}^2}\right)(4 \text{ in} + 4 \text{ in})$$
$$= 12{,}000 \text{ lbf}$$

Solving for the required effective throat thickness,

$$t_e = \frac{12{,}000 \text{ lbf}}{\left(21 \; \frac{\text{kips}}{\text{in}^2}\right)(8 \text{ in})\left(1000 \; \frac{\text{lbf}}{\text{kip}}\right)}$$
$$= 0.07 \text{ in}$$

The nominal weld size is

$$t = \frac{t_e}{0.707} = \frac{0.07 \text{ in}}{0.707}$$
$$= 0.10 \text{ in}$$

3/16

A ⅛ in fillet weld provides adequate capacity.

Check the minimum weld size required based on the thickness of the materials. AISC ASD Table J2.4 specifies the minimum weld size based on the thickness of the thicker part joined.

The thickness of the W10 × 33 flange is found in AISC ASC Part 1.

$$t_f = 0.435 \text{ in}$$

The thickness of the 4 × 4 × ⅜ tube is ⅜ in or 0.375 in. Therefore, the flange thickness controls. The minimum weld size from ASD Table J2.4 is 3/16 in.

Use a 3/16 in fillet weld.

The answer is (B).

Why Other Options Are Wrong

(A) This incorrect solution does not check the minimum weld size based on the thickness of the materials. In this case, the thickness of materials controls the size of the weld. *Uses the nominal tube length in determining weld length.*

(C) This incorrect solution uses the flange thickness of a W10 × 39 column instead of a W10 × 33 column. In this case, the thickness of materials controls the size of the weld.

(D) This incorrect solution does not use the same units for the shear capacity of a weld, R_w, and the allowable shear stress, F_v, when calculating the required weld size.

SOLUTION 57

Check bolts. The applied load per bolt is

$$T \text{ per bolt} = \frac{T}{n} = \frac{60 \text{ kips}}{4 \text{ bolts}}$$
$$= 15.0 \text{ kips/bolt}$$
$$< 26.5 \text{ kips/bolt} \quad [\text{OK}]$$
$$(\text{AISC ASD Table 1-A, Part 4})$$

Use the AISC ASD Preliminary Selection Table for Hangers.

$$b = 2.5 \text{ in}$$

The length of flange, parallel to the leg, tributary to each bolt is

$$p = \frac{10 \text{ in}}{2} = 5.0 \text{ in}$$

The load per inch on the flange is

$$P = \frac{nT}{p} = \frac{(2 \text{ bolts})\left(15.0 \dfrac{\text{kips}}{\text{bolt}}\right)}{5.0 \text{ in}}$$

$$= 6.0 \text{ kips/in}$$

From the AISC ASD Preliminary Selection Table, determine that a $^{15}/_{16}$ in angle is needed.

The smallest 1 in thick angle available is an L6 × 6 × 1. Try an L6 × 6 × 1 double angle.

Using the AISC ASD design method for fasteners loaded in tension, calculate a, b', a', ρ, and δ.

$$a = (\text{length of leg}) - \tfrac{1}{2}t - b$$

$$= 6.0 \text{ in} - \left(\frac{1}{2}\right)(1.0 \text{ in}) - 2.5 \text{ in}$$

$$= 3.0 \text{ in} \quad [2.5\text{ in}]$$

$$a \le 1.25b = (1.25)(2.5 \text{ in})$$

$$= 3.13 \text{ in} \quad [\text{OK}]$$

$$b' = b - \frac{d}{2}$$

$$d = 0.875 \text{ in}$$

$$b' = b - \frac{d}{2} = 2.5 \text{ in} - \frac{0.875 \text{ in}}{2}$$

$$= 2.06 \text{ in}$$

$$a' = a + \frac{d}{2} = 3.0 \text{ in} + \frac{0.875 \text{ in}}{2}$$

$$= 3.44 \text{ in} \quad [2.94]$$

$$\rho = \frac{b'}{a'} = \frac{2.06 \text{ in}}{3.44 \text{ in}}$$

$$= 0.60$$

$$\delta = 1 - \frac{d'}{p}$$

$$d' = \frac{15}{16} \text{ in} = 0.94 \text{ in}$$

$$\delta = 1 - \frac{d'}{p} = 1 - \frac{0.94 \text{ in}}{5.0 \text{ in}}$$

$$= 0.81$$

$$\beta = \left(\frac{1}{\rho}\right)\left(\frac{B}{T} - 1\right)$$

$$= \left(\frac{1}{0.60}\right)\left(\frac{26.5 \text{ kips}}{15.0 \text{ kips}} - 1\right)$$

$$= 1.28 \quad [1.09]$$

Therefore, α' is 1.0.

Calculate the required angle thickness.

$$t_{\text{req}} = \sqrt{\frac{8Tb'}{pF_y(1 + \delta\alpha')}}$$

$$= \sqrt{\frac{(8)(15.0 \text{ kips})(2.06 \text{ in})}{(5.0 \text{ in})\left(36 \dfrac{\text{kips}}{\text{in}^2}\right)(1 + (0.81)(1.0))}}$$

$$= 0.87 \text{ in}$$

$$t_{\text{prov}} = 1.0 \text{ in} > t_{\text{req}} \quad [\text{OK}]$$

Use a double-angle L6 × 6 × 1.

The answer is (C).

Why Other Options Are Wrong

(A) This incorrect solution mistakenly identifies the diameter of the bolt, d, as the leg length and does not check the second-iteration answer.

(B) This incorrect solution calculates the load per inch on the flange considering only one bolt instead of two.

(D) This incorrect solution misreads the allowable tensile load per bolt from AISC ASD Table 1-A for a $^7/_8$ in diameter A490 bolt instead of an A325 bolt and does not check the preliminary size.

SOLUTION 58

A bolt is classified as a dowel type fastener. The allowable load per fastener for dowel type fasteners is given in NDS Table 10.3.1 as

$$Z' = ZC_DC_MC_tC_gC_\Delta C_{eg}C_{di}C_{tn}$$

The diaphragm factor, C_{di}, and the toe-nail factor, C_{tn}, do not apply. The other factors are given as 1.0. Calculate Z and C_g.

NDS Table 11A gives the nominal lateral load for bolts in single shear. Main-member and side-member thickness are both $1^1/_2$ in. For a bolt diameter of $^1/_2$ in, with the load parallel to the grain, and red oak wood

$$Z = 650 \text{ lbf/bolt}$$

From NDS Supplement Table 1B, the areas of the main member and the side member are

$$A_m = A_s = 8.25 \text{ in}^2$$

The spacing between fasteners, s, is 3.25 in. The group action factor, C_g, is calculated according to NDS Sec. 10.3.6. Note that the group action factors listed in NDS Table 10.3.6A are overly conservative in this case and should not be used, since $D < 1$ in and $s < 4$ in. Use NDS Eq. 10.3-1 instead.

$$C_g = \left(\frac{m(1 - m^{2n})}{n\big((1 + R_{\text{EA}}m^n)(1 + m) - 1 + m^{2n}\big)}\right)$$

$$\times \left(\frac{1 + R_{\text{EA}}}{1 - m}\right) \quad [\text{NDS Eq. 10.3-1}]$$

The number of fasteners in a row is defined in NDS Sec. 10.3.6.2. When $a < {}^b/_4$, adjacent rows are considered as one row for the purpose of determining the group action factors.

$$a = 0.75 \text{ in} < \frac{b}{4} = \frac{3.25 \text{ in}}{4}$$

$$= 0.81 \text{ in}$$

Therefore, for the purpose of determining the group action factors, there are two rows of six fasteners in this case.

$$n = 6$$

$$R_{\text{EA}} = \frac{E_s A_s}{E_m A_m}$$

$$= \frac{\left(1.2 \times 10^6 \, \dfrac{\text{lbf}}{\text{in}^2}\right)(8.25 \text{ in}^2)}{\left(1.2 \times 10^6 \, \dfrac{\text{lbf}}{\text{in}^2}\right)(8.25 \text{ in}^2)}$$

$$= 1.0$$

$$\gamma = (180{,}000)D^{1.5}$$

$$= \left(180{,}000 \, \frac{\text{lbf}}{\text{in}}\right)(0.5)^{1.5}$$

$$= 63{,}640 \text{ lbf/in}$$

$$u = 1 + \gamma\left(\frac{s}{2}\right)\left(\frac{1}{E_m A_m} + \frac{1}{E_s A_s}\right)$$

$$= 1 + \left(63{,}640 \, \frac{\text{lbf}}{\text{in}}\right)\left(\frac{3.25 \text{ in}}{2}\right)$$

$$\times \left(\frac{1}{\left(1.2 \times 10^6 \, \dfrac{\text{lbf}}{\text{in}^2}\right)(8.25 \text{ in}^2)} + \frac{1}{\left(1.2 \times 10^6 \, \dfrac{\text{lbf}}{\text{in}^2}\right)(8.25 \text{ in}^2)} \right)$$

$$= 1.02$$

$$m = u - \sqrt{u^2 - 1}$$

$$= 1.02 - \sqrt{(1.02)^2 - 1}$$

$$= 0.82$$

$$C_g = \left(\frac{m(1 - m^{2n})}{n\big((1 + R_{\text{EA}}m^n)(1 + m) - 1 + m^{2n}\big)} \right)$$

$$\times \left(\frac{1 + R_{\text{EA}}}{1 - m} \right)$$

$$= \left(\frac{(0.82)(1 - (0.82)^{(2)(6)})}{(6)\left(\begin{array}{c}(1 + (1.0)(0.82)^6)(1 + 0.82)\\ - 1 + (0.82)^{(2)(6)}\end{array}\right)} \right)$$

$$\times \left(\frac{1 + 1.0}{1 - 0.82} \right)$$

$$= 0.94$$

The capacity of the connection is

$$Z' = Z C_D C_M C_t C_g C_\Delta$$

$$= \left(650 \, \frac{\text{lbf}}{\text{bolt}}\right)(12 \text{ bolts})(1.0)(1.0)(1.0)(0.94)(1.0)$$

$$= 7332 \text{ lbf} \quad (7300 \text{ lbf})$$

The answer is (C).

Why Other Options Are Wrong

(A) This incorrect solution finds the capacity per bolt, not the total capacity of the connection.

(B) This incorrect solution uses the group action factors from NDS Table 10.3.6A. This table is overly conservative in this case because the diameter is less than 1 in and the spacing is less than 4 in, and the footnote to the table indicates that the tabulated values are conservative if even one of the variables is different from those used in the tabulated values.

(D) This incorrect solution does not properly determine the number of fasteners in a row. In this solution, four rows of three fasteners are used. For the purposes of determining the group action factors, NDS Sec. 10.3.6.2 should be used to determine the number of fasteners in a row.

SOLUTION 59

Part 4 of the AISC ASD manual covers connection design. Determine the shear capacity of the bolts, then check for block shear, shear in the angles, bolt bearing on the beam web, and bolt bearing on the connection angles.

Determine the shear capacity of the bolts from AISC ASD Table II-A.

$$L \approx 9 \text{ in} \quad [\text{use } L = 8.5 \text{ in}]$$

$$d = 0.75 \text{ in}$$

The allowable load based on the shear capacity of the bolts is 55.7 kips.

Because the beam is coped, check block shear using AISC ASD Table I-G.

$$l_v = 1.5 \text{ in}$$

$$l_h = 2.0 \text{ in}$$

From the table, determine that C_1 is 1.45 in and C_2 is 0.99 in. The resistance to block shear (allowable load) is

$$R_{\text{BS}} = (C_1 + C_2)F_u t_w$$

From Part I of the AISC ASD manual, determine that for A36 and W14 × 22,

$$F_u = 58 \text{ kips/in}^2$$

$$t_w = 0.23 \text{ in}$$

$$R_{\text{BS}} = (C_1 + C_2)F_u t_w$$

$$= (1.45 \text{ in} + 0.99 \text{ in})\left(58 \, \frac{\text{kips}}{\text{in}^2}\right)(0.23 \text{ in})$$

$$= 32.5 \text{ kips} \quad (33 \text{ kips})$$

Check the shear in the angles using AISC ASD Table II-C.

$$L = 8.5 \text{ in}$$
$$n = 3$$
$$t = {}^5\!/_{16} \text{ in}$$
$$d = {}^3\!/_4 \text{ in}$$

The allowable load is 65.9 kips.

Check the bolt bearing on the beam web using AISC ASD Table I-F.

$$l_v = 1.5 \text{ in}$$
$$F_u = 58 \text{ kips/in}^2$$
$$t_w = 0.23 \text{ in}$$

The allowable load is calculated using the edge distance for each bolt.

$$R = (\text{tabulated value})(\text{no. of bolts})t_w$$

$$= \left(52.2 \, \frac{\text{kips}}{\text{in-bolt}}\right)(3 \text{ bolts})(0.23 \text{ in})$$

$$= 36.0 \text{ kips}$$

Check the bolt bearing on the connection angles using AISC ASD Table I-F.

$$l_v = 1.5 \text{ in}$$
$$F_u = 58 \text{ kips/in}^2$$
$$t_w = {}^5\!/_{16} \text{ in}$$
$$R = (\text{tabulated value})(\text{no. of bolts})t_w$$

$$= \left(52.2 \, \frac{\text{kips}}{\text{in-bolt}}\right)(3 \text{ bolts})(2 \text{ angles})\left(\frac{5}{16} \text{ in}\right)$$

$$= 97.9 \text{ kips}$$

The maximum beam reaction is the smallest of these values. Block shear controls (33 kips).

The answer is (B).

Why Other Options Are Wrong

(A) This incorrect solution does not multiply the bolt bearing values in AISC ASD Table I-F by the number of bolts when calculating the bearing capacities for the beam web and the connection angles.

(C) This incorrect solution uses the wrong AISC ASD Table II-A. It uses the one for A307 or A325 bolts in slip-critical connections instead of that for bearing connections, and it does not check the other conditions.

(D) This incorrect solution bases the allowable load only on the bolt shear and does not check the other conditions.

SOLUTION 60

The weld in this case is subject to both shear and bending. It is customary to design the weld for a uniform shear distribution added vectorially to the maximum bending stress.

For simplicity, the stresses are calculated for a 1 in weld. The required thickness is calculated from the ratio of the actual stress to the weld capacity. Since the weld is on both sides of the angle, the total length of the weld is $2L$.

The shear stress for a 1 in weld is

$$f_s = \frac{P}{A_w} = \frac{P}{2Lw} = \frac{40 \text{ kips}}{(2)(8 \text{ in})(1 \text{ in})}$$

$$= 2.5 \text{ kips/in}^2$$

$$I = \frac{2wL^3}{12} = \frac{(2)(1 \text{ in})(8 \text{ in})^3}{12}$$

$$= 85.33 \text{ in}^4$$

The bending stress is

$$f_b = \frac{Mc}{I} = \frac{(Pe)c}{I} = \frac{(40 \text{ kips})(2.25 \text{ in})(4 \text{ in})}{85.33 \text{ in}^4}$$

$$= 4.22 \text{ kips/in}^2$$

Add vectorially to get the resultant stress.

$$f_r = \sqrt{f_s^2 + f_b^2} = \sqrt{\left(2.5 \, \frac{\text{kips}}{\text{in}^2}\right)^2 + \left(4.22 \, \frac{\text{kips}}{\text{in}^2}\right)^2}$$

$$= 4.90 \text{ kips/in}^2$$

The shear capacity of a fillet weld is

$$R_w = 0.30 t_e F_{u,\text{rod}}$$

For a SMAW fillet weld, t_e is $0.707w$. For E70XX electrodes, $F_{u,\text{rod}}$ is 70 kips/in.

$$R_w = 0.30(0.707w)F_{u,\text{rod}}$$

$$= (0.30)(0.707)w\left(70 \, \frac{\text{kips}}{\text{in}^2}\right)$$

$$= 14.8w \text{ kips/in}^2$$

For a 1 in weld, the resultant stress is

$$f_r = (1 \text{ in})\left(4.90 \, \frac{\text{kips}}{\text{in}^2}\right)$$

$$= 4.90 \text{ kips/in}$$

Equating the resultant stress to the allowable stress,

$$f_r = R_w$$

$$4.90 \frac{\text{kips}}{\text{in}} = 14.8w \frac{\text{kips}}{\text{in}^2}$$

$$w = \frac{4.90 \frac{\text{kips}}{\text{in}}}{14.8 \frac{\text{kips}}{\text{in}^2}} = 0.33 \text{ in}$$

Use a ³⁄₈ in weld.

The answer is (C).

Why Other Options Are Wrong

(A) The weld size is based on shear alone. The bending caused by the eccentricity of the load is ignored in this incorrect solution.

(B) This incorrect solution calculates the weld capacity for the submerged arc weld (SAW) process instead of the shielded metal arc (SMAW) process.

(D) This incorrect solution adds the shear and bending stresses algebraically, not vectorially.

SOLUTION 61

Determine the eccentricity of the load. If the resultant falls outside the column flanges, the anchor bolts must resist the resulting tension.

$$e = \frac{M}{P} = \frac{(200 \text{ ft-kips})\left(12 \frac{\text{in}}{\text{ft}}\right)}{320 \text{ kips}} = 7.5 \text{ in}$$

For a W14 × 109 column,

$$d = 14.32 \text{ in}$$

$$\frac{d}{2} = \frac{14.32 \text{ in}}{2} = 7.16 \text{ in}$$

$$t_f = 0.860 \text{ in}$$

$$\frac{t_f}{2} = \frac{0.860 \text{ in}}{2} = 0.430 \text{ in}$$

$$\frac{d}{2} - \frac{t_f}{2} = 7.16 \text{ in} - 0.43 \text{ in} = 6.73 \text{ in}$$

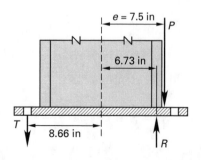

The eccentricity is outside the column flange. Therefore, the bolt must resist uplift. Assume the resultant of the compression forces is located at the center of the column flange. Take moments about this point.

$$\sum M_R = 0 \text{ in-kips} = P\left(e - \left(\frac{d}{2} - \frac{t_f}{2}\right)\right)$$
$$- T\left(\left(\frac{d}{2} + 1.5 \text{ in}\right) + \left(\frac{d}{2} - \frac{t_f}{2}\right)\right)$$
$$0 \text{ in-kips} = (320 \text{ kips})(7.5 \text{ in} - 6.73 \text{ in})$$
$$- T(8.66 \text{ in} + 6.73 \text{ in})$$
$$T = 16.0 \text{ kips}$$

Calculate the size of bolt required. From Table 1-A of the AISC ASD, determine that a 1¹⁄₈ in diameter bolt has an allowable tensile load of 19.9 kips. (A 1 in diameter bolt would be slightly overstressed.)

Alternately, from AISC ASD Table J3.2, determine that the allowable tensile stress for an A307 bolt is 20.0 kips/in². Calculate the required bolt area.

$$A_b = \frac{T}{F_t} = \frac{16.0 \text{ kips}}{20.0 \frac{\text{kips}}{\text{in}^2}}$$
$$= 0.80 \text{ in}^2$$

From AISC ASD Table 1-A, find that the area of a 1¹⁄₈ in diameter bolt is 0.9940 in².

The answer is (D).

Why Other Options Are Wrong

(A) This incorrect solution does not consider the moment in the design. For a base plate without any uplift, a ⁵⁄₈ in diameter bolt is adequate.

(B) This incorrect solution uses the distance to the edge of the column flange instead of the distance to the center of the flange when calculating the tension on the bolt.

(C) This incorrect solution misreads the section properties table and uses the column depth and flange thickness for a W14 × 120 column instead of for a W14 × 109.

SOLUTION 62

Section 11.2.3 of the NDS gives the withdrawal values for a single nail. From NDS Table 11.3.2A, find the specific gravity of southern pine to be 0.55.

NDS Table 11P gives the diameter of a 6d box nail as 0.099 in.

From NDS Table 11.2C, find the tabulated withdrawal design value for a specific gravity of 0.55 and a nail diameter of 0.099 in to be 31 lbf per inch of penetration.

The design withdrawal value is

$$W' = WC_D C_M C_t C_{tn}$$
$$C_D = 1.6 \quad \text{[for wind load combinations]}$$
$$C_M = 1.0 \quad \text{[for moisture content} \le 19\%]$$
$$C_t = 1.0 \quad \text{[according to NDS 10.3.4]}$$
$$C_{tn} = 1.0 \quad \text{[for non-toe-nailed connections]}$$
$$W' = (31\ \text{lbf})(1.6) = 49.6\ \text{lbf}$$

Calculate the maximum area per nail.

$$A = \frac{W'}{w} = \frac{49.6\ \text{lbf}}{36\ \dfrac{\text{lbf}}{\text{ft}^2}}$$
$$= 1.38\ \text{ft}^2$$

If the rafters are spaced at 16 in on center,

$$s_r = (16\ \text{in})\left(\frac{1\ \text{ft}}{12\ \text{in}}\right) = 1.33\ \text{ft}$$

The maximum nail spacing is

$$s_n = \frac{A}{s_r} = \left(\frac{1.38\ \text{ft}^2}{1.33\ \text{ft}}\right)\left(12\ \frac{\text{in}}{\text{ft}}\right)$$
$$= 12.4\ \text{in} \quad (12\ \text{in})$$

The answer is (C).

Why Other Options Are Wrong

(A) This incorrect solution neglects the load duration factor in calculating the withdrawal capacity.

(B) This incorrect solution uses a rafter spacing of 1.5 ft instead of 1.33 ft.

(D) This incorrect solution bases the tabulated withdrawal value on the diameter of a 6d common nail (0.113 in) instead of a 6d box nail.

SOLUTION 63

Calculate the number of bolts per joist.

$$n = \frac{\text{joist spacing}}{\text{bolt spacing}} = \frac{6.0\ \text{ft}}{(16\ \text{in})\left(\dfrac{1\ \text{ft}}{12\ \text{in}}\right)}$$
$$= 4.5\ \text{bolts/joist}$$

Calculate the loads per bolt.

The applied shear load per bolt is

$$b_v = \frac{R}{n} = \frac{3700\ \dfrac{\text{lbf}}{\text{joist}}}{4.5\ \dfrac{\text{bolts}}{\text{joist}}}$$
$$= 822\ \text{lbf/bolt}$$

The prying tension is

$$T = \frac{Rx}{y} = \frac{(3700\ \text{lbf})(2.5\ \text{in})}{3\ \text{in}}$$
$$= 3083\ \text{lbf/joist}$$

The prying tension per bolt is

$$b_a = \frac{T}{n} = \frac{3083\ \dfrac{\text{lbf}}{\text{joist}}}{4.5\ \dfrac{\text{bolts}}{\text{joist}}}$$
$$= 685\ \text{lbf/bolt}$$

Determine the allowable loads using the MSJC code. The allowable load in shear is the lesser of the following two options.

$$B_v = 350\sqrt[4]{f'_m A_b} \quad \text{[MSJC Eq. 2-5]}$$
$$B_v = 0.12 A_b f_y \quad \text{[MSJC Eq. 2-6]}$$

Try a 3/4 in diameter bolt.

$$A_b = 0.44\ \text{in}^2$$
$$B_v = 350\sqrt[4]{f'_m A_b} = (350)\left(\sqrt[4]{(1500)(0.44)}\ \frac{\text{lbf}}{\text{in}^2}(\text{in}^2)\right)$$
$$= 1774\ \text{lbf} \quad \text{[The masonry controls.]}$$
$$B_v = 0.12 A_b f_y = (0.12)(0.44\ \text{in}^2)\left(36{,}000\ \frac{\text{lbf}}{\text{in}^2}\right)$$
$$= 1901\ \text{lbf}$$

The allowable load in tension is the lesser of

$$B_a = 0.5 A_p \sqrt{f'_m} \quad \text{[MSJC Eq. 2-1]}$$
$$B_a = 0.2 A_b f_y \quad \text{[MSJC Eq. 2-2]}$$
$$A_p = \pi l_b^2 = \pi(6\ \text{in})^2$$
$$= 113\ \text{in}^2 \quad \text{[MSJC Eq. 2-3]}$$
$$B_a = 0.5 A_p \sqrt{f'_m} = (0.5)(113\ \text{in}^2)\left(\sqrt{1500}\frac{\text{lbf}}{\text{in}^2}\right)$$
$$= 2190\ \text{lbf} \quad \text{[The masonry controls.]}$$
$$B_a = 0.2 A_b f_y = (0.2)(0.44\ \text{in}^2)\left(36{,}000\ \frac{\text{lbf}}{\text{in}^2}\right)$$
$$= 3168\ \text{lbf}$$

Check combined shear and tension using the interaction equation.

$$\frac{b_a}{B_a} + \frac{b_v}{B_v} \le 1 \quad \text{[MSJC Eq. 2-7]}$$
$$\frac{685\ \text{lbf}}{2190\ \text{lbf}} + \frac{822\ \text{lbf}}{1774\ \text{lbf}} = 0.78 \quad \text{[OK]}$$

However, since this result is understressed and the problem asks for the smallest bolt size, try a smaller size.

Try a 5/8 in diameter bolt.

$$A_b = 0.31 \text{ in}^2$$

The allowable shear load is the lesser of

$$B_v = 350 \sqrt[4]{f'_m A_b} = (350) \left(\sqrt[4]{(1500)(0.31)} \frac{\text{lbf}}{\text{in}^2}(\text{in}^2) \right)$$

$$= 1625 \text{ lbf}$$

$$B_v = 0.12 A_b f_y = (0.12)(0.31 \text{ in}^2) \left(36{,}000 \frac{\text{lbf}}{\text{in}^2} \right)$$

$$= 1339 \text{ lbf} \quad \text{[The bolt steel controls.]}$$

The allowable load in tension is the lesser of

$$B_a = 0.5 A_p \sqrt{f'_m} = (0.5)(113 \text{ in}^2) \left(\sqrt{1500} \frac{\text{lbf}}{\text{in}^2} \right)$$

$$= 2190 \text{ lbf} \quad \text{[The masonry controls.]}$$

$$B_a = 0.2 A_b f_y = (0.2)(0.31 \text{ in}^2) \left(36{,}000 \frac{\text{lbf}}{\text{in}^2} \right)$$

$$= 2232 \text{ lbf}$$

Check combined shear and tension using the interaction equation.

$$\frac{b_a}{B_a} + \frac{b_v}{B_v} \leq 1 \quad \text{[MSJC Eq. 2-7]}$$

$$\frac{685 \text{ lbf}}{2190 \text{ lbf}} + \frac{822 \text{ lbf}}{1339 \text{ lbf}} = 0.93 \quad \text{[OK]}$$

Since this result is near the combined stress limit, use 5/8 in diameter bolts.

The answer is (C).

Why Other Options Are Wrong

(A) This incorrect solution mistakes the load *from* the joist as the total load *on* the joist and divides the load in half to get the resultant. It is also incorrect in only considering the shear load.

(B) This incorrect solution improperly calculates the load per bolt. The bolt spacing is ignored, and the loads per foot are calculated instead of the loads per bolt. This answer is slightly overstressed but OK.

(D) This incorrect solution calculates the stresses correctly but stops with a 3/4 in diameter bolt and does not find the smallest size bolt.

SOLUTION 64

NDS Sec. 15.3.2 contains the equations for calculating the column stability factor for built-up columns. The column stability factor is based on the slenderness ratios

and must be calculated in each direction. The smaller value is used to determine the allowable compression design value parallel to the grain, F'_c, for the column.

First, determine the dimensional properties of the column.

From NDS Supplement Table 1B, determine the actual dimensions of a 2×6 sawn lumber to be 1.5 in by 5.5 in.

Use NDS Fig. 15B to determine that

$$d_1 = 5.5 \text{ in}$$
$$d_2 = (3)(1.5 \text{ in}) = 4.5 \text{ in}$$
$$l_1 = (9 \text{ ft}) \left(12 \frac{\text{in}}{\text{ft}} \right) = 108 \text{ in}$$
$$l_2 = (4.5 \text{ ft}) \left(12 \frac{\text{in}}{\text{ft}} \right) = 54 \text{ in}$$

The column stability factor, C_p, is based on the effective column length, l_e.

$$l_e = K_e l \quad \text{[NDS Sec. 15.3.2.1]}$$

From Table G1 in App. G of the NDS, determine that the buckling length coefficient, K_e, is 1.0 for a column with both ends free to rotate but not free to translate.

$$l_{e1} = K_e l_1 = (1.0)(108 \text{ in})$$
$$= 108 \text{ in}$$
$$l_{e2} = K_e l_2 = (1.0)(54 \text{ in})$$
$$= 54 \text{ in}$$

Calculate the slenderness ratios. NDS Sec. 15.3.2.3 specifies that the slenderness ratios shall not exceed 50. The slenderness ratios are

$$\frac{l_{e1}}{d_1} = \frac{108 \text{ in}}{5.5 \text{ in}} = 19.6 < 50 \quad \text{[OK]}$$

$$\frac{l_{e2}}{d_2} = \frac{54 \text{ in}}{4.5 \text{ in}} = 12.0 < 50 \quad \text{[OK]}$$

Calculate the column stability factor in each direction.

$$C_p = K_f \left(\frac{1 + \frac{F_{c,E}}{F_c^*}}{2c} - \sqrt{\left(\frac{1 + \frac{F_{c,E}}{F_c^*}}{2c} \right)^2 - \frac{F_{c,E}}{F_c^*}} \right)$$

$$\text{[NDS Eq. 15.3-1]}$$

$$F_c^* = 1750 \text{ lbf/in}^2 \quad \text{[given]}$$

$$F_{c,E} = \frac{K_{c,E} E'}{\left(\frac{l_e}{d} \right)^2}$$

$$K_{c,E} = 0.3 \quad \text{[given]}$$

$$E' = 1.7 \times 10^6 \text{ lbf/in}^2 \quad \text{[given]}$$

$$c = 0.8 \quad \text{[for sawn lumber]}$$

Direction 1:

$$\frac{l_{e1}}{d_1} = 19.6$$

$$K_f = 1.0 \quad \left[\begin{array}{c} \text{for a built-up column where } l_{e1}/d_1 \\ \text{is used to calculate } F_{c,E} \end{array} \right]$$

$$K_{c,E} = 0.3$$

$$F_{c,E_1} = \frac{K_{c,E}E'}{\left(\frac{l_{e1}}{d_1}\right)^2} = \frac{(0.3)\left(1.7 \times 10^6 \, \frac{\text{lbf}}{\text{in}^2}\right)}{(19.6)^2}$$

$$= 1328 \, \text{lbf/in}^2$$

$$\frac{F_{c,E_1}}{F_c^*} = \frac{1328 \, \frac{\text{lbf}}{\text{in}^2}}{1750 \, \frac{\text{lbf}}{\text{in}^2}} = 0.759$$

$$C_{p1} = (1.0)$$

$$\times \left(\frac{1 + 0.759}{(2)(0.8)} - \sqrt{\left(\frac{1 + 0.759}{(2)(0.8)}\right)^2 - \frac{0.759}{0.8}} \right)$$

$$= 0.59$$

Direction 2:

$$\frac{l_{e2}}{d_2} = 12.0$$

$$K_f = 0.6 \quad \left[\begin{array}{c} \text{for a built-up column where } l_{e2}/d_2 \text{ is used} \\ \text{to calculate } F_{c,E} \text{ and the column is nailed} \end{array} \right]$$

$$K_{c,E} = 0.3$$

$$F_{c,E_2} = \frac{K_{c,E}E'}{\left(\frac{l_{e2}}{d_2}\right)^2} = \frac{(0.3)\left(1.7 \times 10^6 \, \frac{\text{lbf}}{\text{in}^2}\right)}{(12.0)^2}$$

$$= 3542 \, \text{lbf/in}^2$$

$$\frac{F_{c,E_2}}{F_c^*} = \frac{3542 \, \frac{\text{lbf}}{\text{in}^2}}{1750 \, \frac{\text{lbf}}{\text{in}^2}} = 2.0$$

$$C_{p2} = (0.6) \left(\frac{1 + 2.0}{(2)(0.8)} - \sqrt{\left(\frac{1 + 2.0}{(2)(0.8)}\right)^2 - \frac{2.0}{0.8}} \right)$$

$$= 0.52$$

The critical column stability factor is the smaller of C_{p1} or C_{p2}, which is 0.52.

The answer is (B).

Why Other Options Are Wrong

(A) This incorrect solution reverses the dimensions of the column, d_1 and d_2, in calculating the slenderness ratios.

(C) This incorrect solution uses a buckling length coefficient, K_e, of 0.65 from NDS Table G1 in App. G for a column with rotation and translation fixed at both ends.

(D) This incorrect solution uses the K_{f2} for bolted columns instead of nailed columns. It could also be arrived at by calculating C_p correctly but selecting the larger value as the critical one.

SOLUTION 65

Interaction diagrams such as those found in the *Civil Engineering Reference Manual* or other sources provide a means to determine the required area of steel when the loads and section properties of a column are known.

$$\frac{\phi P_n}{A_g} = \frac{P_u}{A_g} = \frac{750 \, \text{kips}}{(20 \, \text{in})(24 \, \text{in})}$$

$$= 1.56 \, \text{kips/in}^2$$

$$\frac{\phi M_n}{A_g h} = \frac{M_u}{A_g h} = \frac{(600 \, \text{ft-kips})\left(12 \, \frac{\text{in}}{\text{ft}}\right)}{(20 \, \text{in})(24 \, \text{in})(24 \, \text{in})}$$

$$= 0.63 \, \text{kip/in}^2$$

From the relevant interaction diagram (in the *Civil Engineering Reference Manual*), determine that ρ_g is 0.03.

$$A_s = \rho_g A_g = (0.03)(20 \, \text{in})(24 \, \text{in})$$

$$= 14.4 \, \text{in}^2 \quad (14 \, \text{in}^2)$$

The answer is (B).

Why Other Options Are Wrong

(A) This incorrect solution uses the interaction diagram for $\gamma = 0.90$ (in the *Civil Engineering Reference Manual*).

(C) This incorrect solution uses the interaction diagram for columns with steel distributed evenly on four faces (in the *Civil Engineering Reference Manual*) instead of the diagram for steel on only two faces.

(D) This solution incorrectly calculates the moment factor for use with the interaction diagram. The units do not work out.

$$\frac{\phi M_n}{A_g h} = \frac{M_u}{A_g h} = \frac{600 \, \text{ft-kips}}{(20 \, \text{in})(24 \, \text{in})}$$

$$= 1.25 \, \text{ft-kips/in}^2$$

SOLUTION 66

When slenderness must be considered in the design of compression members, the magnified moment procedure can be used if a more refined analysis is not performed. According to Sec. 10.11.5 of ACI 318, if $kl_u/r < 100$, magnified moments can be used.

ACI 318 Sec. 10.12 contains the provisions for magnified moments in nonsway frames. According to ACI 318 Sec. 10.12.2, if $kl_u/r \leq 34 - 12(M_1/M_2)$ and is no more than 40, slenderness can be ignored.

The unsupported length of a compression member is taken as the clear distance between floor slabs.

$$l_u = (13.0 \text{ ft})\left(12 \, \frac{\text{in}}{\text{ft}}\right) - 6 \text{ in}$$
$$= 150 \text{ in}$$

For a 12 in diameter column,

$$r = \frac{d}{4} = \frac{12 \text{ in}}{4}$$
$$= 3 \text{ in}$$
$$I_g = \frac{\pi d^4}{64} = \frac{\pi (12 \text{ in})^4}{64}$$
$$= 1018 \text{ in}^4$$
$$k = 1.0 \quad [\text{ACI 318 Sec. 10.12.1}]$$
$$\frac{kl_u}{r} = \frac{(1.0)(150 \text{ in})}{3 \text{ in}} = 50$$
$$M_1 = -100 \text{ ft-kips} \quad \left[\begin{array}{l}\text{for columns bent in}\\\text{double curvature}\end{array}\right]$$
$$M_2 = 100 \text{ ft-kips}$$
$$34 - 12\left(\frac{M_1}{M_2}\right) = 34 - (12)\left(\frac{-100 \text{ ft-kips}}{100 \text{ ft-kips}}\right)$$
$$= 46 \quad [\text{so } 40 < kl_u/r < 100]$$

Therefore, slenderness must be considered and magnified moments can be used.

The magnified moment is given by ACI 318 Eq. 10-8.

$$M_c = \delta_{ns} M_2$$
$$\delta_{ns} = \frac{C_m}{1 - \dfrac{P_u}{0.75 P_c}} \geq 1.0 \quad [\text{ACI 318 Eq. 10-9}]$$
$$P_c = \frac{\pi^2 EI}{(kl_u)^2} \quad [\text{ACI 318 Eq. 10-10}]$$
$$EI = \frac{0.4 E_c I_g}{1 + \beta_d} \quad [\text{ACI 318 Eq. 10-12}]$$

The ratio of the maximum factored axial dead load to the total factored axial load, β_d, can be taken as 0.6 according to ACI 318 Sec. R10.12.3.

$$EI = \frac{0.4 E_c I_g}{1 + \beta_d} = \frac{(0.4)\left(3.6 \times 10^6 \, \frac{\text{lbf}}{\text{in}^2}\right)(1018 \text{ in}^4)}{1 + 0.6}$$
$$= 9.16 \times 10^8 \text{ lbf-in}^2$$
$$P_c = \frac{\pi^2 EI}{(kl_u)^2} = \frac{\pi^2 (9.16 \times 10^8 \text{ lbf-in}^2)\left(\dfrac{1 \text{ kip}}{1000 \text{ lbf}}\right)}{((1.0)(150 \text{ in}))^2}$$
$$= 401.8 \text{ kips}$$

$$C_m = 0.6 + 0.4\left(\frac{M_1}{M_2}\right) \geq 0.4 \quad [\text{ACI 318 Eq. 10-13}]$$
$$C_m = 0.6 + (0.4)\left(\frac{-100 \text{ ft-kips}}{100 \text{ ft-kips}}\right)$$
$$= 0.2 \quad [\text{use } 0.4]$$
$$\delta_{ns} = \frac{C_m}{1 - \dfrac{P_u}{0.75 P_c}} = \frac{0.4}{1 - \dfrac{300 \text{ kips}}{(0.75)(401.8 \text{ kips})}}$$
$$= 89.3 \geq 1.0 \quad [\text{OK}]$$
$$M_c = \delta_{ns} M_2 = (89.3)(100 \text{ ft-kips})$$
$$= 8930 \text{ ft-kips} \quad (9000 \text{ ft-kips})$$

It is interesting to note that even for kl_u/r just over 40 (50 in this case), the moment magnifier, δ_{ns}, is large.

The answer is (C).

Why Other Options Are Wrong

(A) This incorrect solution uses the wrong values for EI in ACI 318 Eq. 10-10 for the critical load, P_c. EI as used in ACI 318 Eq. 10-10 is given by Eq. 10-11 or Eq. 10-12. This wrong solution uses $E_c I$.

(B) This incorrect solution ignores the lower limit for C_m in ACI 318 Eq. 10-13.

(D) This incorrect solution uses a positive sign for the M_1/M_2 ratio. A positive sign indicates single curvature, not double.

SOLUTION 67

The beams framing into the first-floor corner column are exterior beams and have flanges on only one side. Section 13.2.4 of ACI 318 defines a beam in a two-way slab system as including that portion of the slab on each side of the beam, b_f, extending a distance equal to the projection of the beam above or below the slab, whichever is greater, but not greater than four times the slab thickness.

$$b_f = 14 \text{ in} \leq 4t = (4)(6 \text{ in})$$
$$= 24 \text{ in}$$

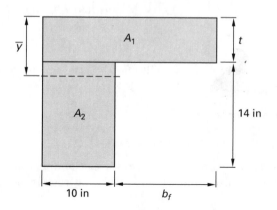

Compute I_b.

	A (in^2)	y (in)	Ay (in^3)	$A(y-\bar{y})^2$ (in^4)	I_o (in^4)
A_1	144	3	432	3500	432
A_2	140	13	1820	3599	2287
totals	284		2252	7099	2719

$$\bar{y} = \frac{\sum Ay}{\sum A} = \frac{2252 \text{ in}^3}{284 \text{ in}^2}$$
$$= 7.93 \text{ in}$$
$$I_b = I_o + A(y-\bar{y})^2$$
$$= 2719 \text{ in}^4 + 7099 \text{ in}^4$$
$$= 9818 \text{ in}^4$$

According to ACI 318 Sec. R10.12, when calculating Ψ, for flexural members,

$$I_{b,e} = 0.35 I_b = (0.35)(9818 \text{ in}^4)$$
$$= 3436 \text{ in}^4 \quad (3400 \text{ in}^4)$$

The answer is (A).

Why Other Options Are Wrong

(B) This incorrect solution uses the definition of a T-beam found in ACI 318 Sec. 8.10.3 when determining the extent beam flange. This approach is applicable to one-way slabs only.

(C) In this incorrect solution, the gross moment of inertia is calculated instead of I_e.

(D) This incorrect solution calculates the beam flange correctly but does not recognize the beam as being an exterior beam with a flange on only one side.

SOLUTION 68

The relative stiffness parameter is the same in each direction for a corner column.

$$\Psi_{\text{E-W}} = \Psi_{\text{N-S}}$$
$$= \frac{\sum \left(\frac{EI}{l_c}\right)_{\text{compr}}}{\sum \left(\frac{EI}{l_c}\right)_{\text{flex}}} \quad \left[\begin{array}{c} \text{ACI 318} \\ \text{Fig. R10.12.1} \end{array}\right]$$

Since E is the same throughout,

$$\Psi = \frac{\sum \left(\frac{I}{l_c}\right)_{\text{compr}}}{\sum \left(\frac{I}{l_c}\right)_{\text{flex}}}$$
$$I_c = \frac{bh^3}{12} = \frac{(18 \text{ in})^4}{12}$$
$$= 8748 \text{ in}^4$$

$$I_b = 10{,}000 \text{ in}^4$$

Adjust for cracking and creep (ACI 318 Sec. R10.12.1).

$$I_{c,e} = 0.70 I_c$$
$$= (0.70)(8748 \text{ in}^4)$$
$$= 6124 \text{ in}^4$$

$$I_{b,e} = 0.35 I_b$$
$$= (0.35)(10{,}000 \text{ in}^4)$$
$$= 3500 \text{ in}^4$$

$$\Psi_{\text{top}} = \frac{\dfrac{6124 \text{ in}^4}{13 \text{ ft}} + \dfrac{6124 \text{ in}^4}{18 \text{ ft}}}{\dfrac{3500 \text{ in}^4}{20 \text{ ft}}}$$

$$= 4.64 \quad (4.6)$$

Because the column is a corner column, there is only one beam in each direction.

The answer is (B).

Why Other Options Are Wrong

(A) This incorrect solution calculates the relative stiffness parameter for an interior column, not a corner column.

(C) This incorrect solution mistakenly uses 13 ft instead of 18 ft for the unbraced column length on the first story.

(D) This incorrect solution does not adjust the moment of inertia for cracking and creep as recommended in ACI 318 Sec. R10.12.1.

SOLUTION 69

Section 10.12 of ACI 318 contains a nomograph to determine the effective length factor, k, knowing the relative stiffness parameters.

$$\Psi = \frac{\sum \left(\frac{EI}{l_c}\right)_{\text{compr}}}{\sum \left(\frac{EI}{l_c}\right)_{\text{flex}}} \quad [\text{ACI 318 Fig. R10.12.1}]$$

Since the modulus of elasticity is the same throughout,

$$\Psi = \frac{\sum \left(\frac{I}{l_c}\right)_{\text{compr}}}{\sum \left(\frac{I}{l_c}\right)_{\text{flex}}}$$
$$I_c = 8800 \text{ in}^4$$
$$I_b = 10{,}000 \text{ in}^4$$

Adjust for cracking and creep (ACI 318 Sec. R10.12.1).

$$I_{c,e} = 0.70I_c = (0.70)(8800 \text{ in}^4)$$
$$= 6160 \text{ in}^4$$

$$I_{b,e} = 0.35I_b = (0.35)(10,000 \text{ in}^4)$$
$$= 3500 \text{ in}^4$$

$$\Psi_{\text{top}} = \frac{\dfrac{6160 \text{ in}^4}{13 \text{ ft}} + \dfrac{6160 \text{ in}^4}{13 \text{ ft}}}{\dfrac{3500 \text{ in}^4}{20 \text{ ft}} + \dfrac{3500 \text{ in}^4}{20 \text{ ft}}}$$
$$= 2.71$$

$$\Psi_{\text{bot}} = \frac{\dfrac{6160 \text{ in}^4}{13 \text{ ft}} + \dfrac{6160 \text{ in}^4}{18 \text{ ft}}}{\dfrac{3500 \text{ in}^4}{20 \text{ ft}} + \dfrac{3500 \text{ in}^4}{20 \text{ ft}}}$$
$$= 2.33$$

Using the nomograph in ACI 318 Fig. R10.12.1(b) for unbraced frames, determine the effective length factor in the East/West direction to be 1.71.

The effective length of the interior second-story column in the East/West direction is

$$l_e = kl_u = (1.71)(\cancel{13} \overset{12.5}{} \text{ ft})$$
$$= \cancel{22.2} \text{ ft} \quad (\cancel{22} \text{ ft})$$
$$ 21.4 \quad 21$$

The answer is (C).

Why Other Options Are Wrong

(A) In this incorrect solution, the relative stiffness parameter is inverted and the nomograph for braced frames is used.

(B) This incorrect solution does not adjust the moment of inertia for cracking and creep.

(D) This incorrect solution mistakenly uses 13 ft instead of 18 ft for the unbraced column length on the first story.

SOLUTION 70

Use the following equation for calculating the maximum biaxial load.

$$\frac{1}{P_{\text{biaxial}}} = \frac{1}{P_x} + \frac{1}{P_y} - \frac{1}{P_o}$$

Use column interaction diagrams (such as those in App. 70 of the *Civil Engineering Reference Manual* or a similar reference) to determine the maximum axial load in each direction.

In the x direction,

$$e_x = 5.0 \text{ in}$$

The section properties of the column are based on the actual column dimensions. For a 20 in brick masonry column, the actual dimensions are

$$b = t = 19.625 \text{ in}$$
$$A_s = \text{four no. 6 bars} = (4)(0.44 \text{ in}^2)$$
$$= 1.76 \text{ in}^2$$

$$\rho_t = \frac{A_s}{bt} = \frac{1.76 \text{ in}^2}{(19.625 \text{ in})(19.625 \text{ in})}$$
$$= 0.0046$$

$$E_m = 700f'_m \quad \text{[for clay masonry]}$$
$$= (700)\left(3500 \frac{\text{lbf}}{\text{in}^2}\right)$$
$$= 2.45 \times 10^6 \text{ lbf/in}^2$$

$$E_s = 29 \times 10^6 \text{ lbf/in}^2$$

$$n = \frac{E_s}{E_m} = \frac{29 \times 10^6 \dfrac{\text{lbf}}{\text{in}^2}}{2.45 \times 10^6 \dfrac{\text{lbf}}{\text{in}^2}}$$
$$= 11.8$$

$$n\rho_t = (11.8)(0.0046)$$
$$= 0.054 \quad \text{[use 0.05]}$$

$$\frac{e_x}{t} = \frac{5.0 \text{ in}}{19.625 \text{ in}}$$
$$= 0.255$$

From column interaction diagrams, $P/F_b bt$ is 0.38.

$$P_x = 0.38F_b bt = (0.38)\tfrac{1}{3}f'_m bt$$
$$= (0.38)\left(\frac{1}{3}\right)\left(3500 \frac{\text{lbf}}{\text{in}^2}\right)(19.625 \text{ in})(19.625 \text{ in})$$
$$= 170,746 \text{ lbf}$$

In the y direction, e is 6.2 in.

$$\frac{e_y}{t} = \frac{6.2 \text{ in}}{19.625 \text{ in}} = 0.316$$

$$n\rho_t = (11.8)(0.0046)$$
$$= 0.054 \quad \text{[use 0.05]}$$

From column interaction diagrams, $P/F_b bt$ is 0.26.

$$P_y = 0.26F_b bt = (0.26)\left(\tfrac{1}{3}\right)f'_m bt$$
$$= (0.26)\left(\frac{1}{3}\right)\left(3500 \frac{\text{lbf}}{\text{in}^2}\right)(19.625 \text{ in})(19.625 \text{ in})$$
$$= 116,826 \text{ lbf}$$

Find P_o. Since $h/r < 99$, use Eq. 2-17 from the MSJC code. The contribution from the steel is relatively small. Ignoring the steel,

$$P_o = (0.25f'_m A_n)\left(1 - \left(\frac{h}{140r}\right)^2\right)$$

$$= \left((0.25)\left(3500 \ \frac{\text{lbf}}{\text{in}^2}\right)\right.$$

$$\left. \times (19.625 \text{ in})(19.625 \text{ in})\right)\left(1 - \left(\frac{72}{140}\right)^2\right)$$

$$= 247{,}866 \text{ lbf}$$

$$\frac{1}{P_{\text{biaxial}}} = \frac{1}{P_x} + \frac{1}{P_y} - \frac{1}{P_o}$$

$$= \frac{1}{170{,}746 \text{ lbf}} + \frac{1}{116{,}826 \text{ lbf}} - \frac{1}{247{,}866 \text{ lbf}}$$

$$P_{\text{biaxial}} = 96{,}321 \text{ lbf} \quad (96{,}000 \text{ lbf})$$

The answer is (B).

Why Other Options Are Wrong

(A) This incorrect solution makes an addition error in the biaxial equation, adding the inverse of the pure axial load rather than subtracting it.

(C) This incorrect solution uses the diameter of a no. 6 bar as its area when calculating A_s.

(D) This incorrect solution calculates the maximum axial load for zero eccentricity using MSJC code Eq. 2-17.

SOLUTION 71

Determine the factored loads on the columns.

$$P_u = \overset{1\cdot 2}{\cancel{1.4}} P_D + \overset{1\cdot 6}{\cancel{1.7}} P_L$$

$$P_{u1} = (1.4)(100 \text{ kips}) + (1.7)(60 \text{ kips})$$
$$= 242 \text{ kips}$$
$$P_{u2} = (1.4)(60 \text{ kips}) + (1.7)(30 \text{ kips})$$
$$= 135 \text{ kips}$$

The ultimate soil pressure is

$$q_u = \frac{P_{u1} + P_{u2}}{BL}$$

The footing width and length are

$$B = 4.5 \text{ ft}$$
$$L = 20 \text{ ft}$$
$$q_u = \frac{242 \text{ kips} + 135 \text{ kips}}{(4.5 \text{ ft})(20 \text{ ft})}$$
$$= 4.19 \text{ kips/ft}^2$$

The design moment at the face of column 1 in the longitudinal direction is

$$M_u = \frac{q_u B x^2}{2} = \frac{\left(4.19 \ \frac{\text{kips}}{\text{ft}^2}\right)(4.5 \text{ ft})(5 \text{ ft})^2}{2}$$

$$= 236 \text{ ft-kips}$$

The required reinforcement can be determined easily by utilizing design aids such as Graph A.1a in *Design of Concrete Structures* or equations as shown in the *Civil Engineering Reference Manual*.

Using Graph A.1a in *Design of Concrete Structures*, enter the graph knowing

$$f'_c = 3000 \text{ lbf/in}^2$$
$$f_y = 60{,}000 \text{ lbf/in}^2$$
$$\phi = 0.9 \quad \text{[for flexure]}$$
$$R = \frac{M_u}{\phi b d^2}$$

$$= \frac{(236 \text{ ft-kips})\left(1000 \ \frac{\text{lbf}}{\text{kip}}\right)}{(0.9)(4.5 \text{ ft})(15 \text{ in})^2}$$

$$= 259 \text{ lbf/in}^2$$

Determine that

$$\rho = 0.004$$
$$A_s = \rho b d$$

$$= (0.004)(4.5 \text{ ft})\left(12 \ \frac{\text{in}}{\text{ft}}\right)(15 \text{ in})$$

$$= 3.24 \text{ in}^2$$

Eight no. 6 bars provide 3.53 in^2 of steel.

The answer is (D).

Why Other Options Are Wrong

(A) This incorrect solution calculates the design moment and the area of steel required for a 12 in unit width and not for the entire width.

(B) This incorrect solution uses service loads instead of factored loads. Service loads are used to size the footing, but design of the footing is based on factored loads.

(C) This incorrect solution assumes that six bars are to be used (instead of no. 6 bars) in determining the area of reinforcement required.

SOLUTION 72

Calculate the soil pressure on the footing.

$$q_u = \frac{P_u}{BL} = \frac{60 \text{ kips}}{(10 \text{ ft})(5 \text{ ft})}$$

$$= 1.2 \text{ kips/ft}^2$$

The critical section for the design of the footing reinforcement is at the face of the column.

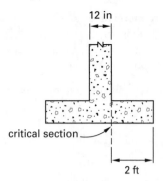

12 in

critical section

2 ft

The moment acting at the critical section from a uniformly distributed soil pressure is

$$M_u = \frac{q_u B b^2}{2} = \frac{\left(1.2 \frac{\text{kips}}{\text{ft}}\right)(10 \text{ ft})(2 \text{ ft})^2}{2}$$

$$= 24.0 \text{ ft-kips}$$

The required area of steel is

$$A_s = \frac{M_u}{\phi f_y (d - \lambda)}$$

To begin, estimate λ as $0.1d$.

$$A_s = \frac{M_u}{\phi f_y (d - \lambda)} = \frac{(24.0 \text{ ft-kips})\left(12 \frac{\text{in}}{\text{ft}}\right)}{(0.9)\left(60 \frac{\text{kips}}{\text{in}^2}\right)(12 \text{ in} - 1.2 \text{ in})}$$

$$= 0.49 \text{ in}^2$$

Check the assumed λ value.

$$A_c = \frac{f_y A_s}{0.85 f'_c} = \frac{\left(60 \frac{\text{kips}}{\text{in}^2}\right)(0.49 \text{ in}^2)}{(0.85)\left(3 \frac{\text{kips}}{\text{in}^2}\right)}$$

$$= 11.5 \text{ in}^2$$

The compression zone is rectangular with a width of

$$B = (10 \text{ ft})\left(12 \frac{\text{in}}{\text{ft}}\right) = 120 \text{ in}$$

Therefore,

$$\lambda = \frac{A_c}{2B} = \frac{11.5 \text{ in}^2}{(2)(120 \text{ in})}$$

$$= 0.048 \text{ in}$$

The revised area of steel is

$$A_s = \frac{M_u}{\phi f_y (d - \lambda)}$$

$$= \frac{(24.0 \text{ ft-kips})\left(12 \frac{\text{in}}{\text{ft}}\right)}{(0.9)\left(60 \frac{\text{kips}}{\text{in}^2}\right)(12 \text{ in} - 0.048 \text{ in})}$$

$$= 0.45 \text{ in}^2$$

The minimum area of steel is

$$A_{s,\min} = \rho_{\min} B d = (0.0018)(120 \text{ in})(12 \text{ in})$$

$$= 2.59 \text{ in}^2 \quad [\text{ACI 318 Sec. 7.12.2.1}]$$

The total area of steel in the short direction is

$$A_{\text{sd}} = 2.59 \text{ in}^2$$

In a two-way rectangular footing, the reinforcement in the short direction is not distributed uniformly. Instead, a portion of the total reinforcement is distributed uniformly over a band centered on the column centerline equal to the length of the short side of the footing.

$$A_1 = A_{\text{sd}}\left(\frac{2}{\beta + 1}\right) \quad [\text{ACI 318 Eq. 15-1}]$$

$$A_1 = (2.59 \text{ in}^2)\left(\frac{2}{2 + 1}\right)$$

$$= 1.73 \text{ in}^2$$

The remainder is distributed uniformly in the remaining part of the footing.

The reinforcement under the column is 1.73 in². Nine no. 4 bars provides 1.77 in².

Check the maximum spacing of the bars. ACI 318 Sec. 10.5.4 limits the spacing of reinforcement in a footing to 18 in. Nine no. 4 bars spaced over 60 in is less than 18 in maximum spacing.

The answer is (C).

Why Other Options Are Wrong

(A) This incorrect solution calculates the reinforcement required under the column in the long direction of the footing instead of in the short direction.

(B) This incorrect solution calculates the reinforcement required in the long direction of the footing and does not differentiate between the total reinforcement required and that under the column.

(D) This incorrect solution assumes that the minimum reinforcement is all located under the column, and it does not distribute the reinforcement according to ACI 318 Sec. 15.4.4.2.

SOLUTION 73

The necessary equations for calculating the stresses in masonry can be found in a masonry handbook and in the MSJC code.

The allowable moment based on flexural compressive stress in the masonry is

$$M_m = \frac{1}{2} F_b b d^2 j k$$

$$F_b = \frac{1}{3} f'_m \quad [\text{MSJC Sec. 2.3.3}]$$

The allowable moment based on flexural tensile stress in the steel is

$$M_s = A_s F_s j d$$

For Grade 60 reinforcement,

$$F_s = 24{,}000 \text{ lbf/in}^2 \quad \text{[MSJC Sec. 2.3.2]}$$

$$\rho = \frac{A_s}{bd}$$
$$k = \sqrt{2\rho n + (\rho n)^2} - \rho n$$
$$j = 1 - \frac{k}{3}$$
$$n = \frac{E_s}{E_m} = 21.5 \quad \text{[given]}$$

Assume the vertical reinforcement is centered in the middle of the wall. The effective depth to reinforcement is

$$d = \frac{t}{2} = \frac{5.63 \text{ in}}{2} = 2.8 \text{ in}$$

Determine the required area of reinforcement,

$$A_{s,\text{req}} = \frac{M}{F_s j d}$$

The maximum moment on the wall is

$$M_{\max} = \left(341 \; \frac{\text{ft-lbf}}{\text{ft}}\right)\left(12 \; \frac{\text{in}}{\text{ft}}\right) = 4092 \text{ in-lbf/ft}$$

Estimate that j is equal to 0.9 and use the $\frac{1}{3}$ allowable stress increase for wind.

$$A_{s,\text{req}} = \frac{M_{\max}}{F_s j d}$$
$$= \frac{4092 \; \frac{\text{in-lbf}}{\text{ft}}}{\left(24{,}000 \; \frac{\text{lbf}}{\text{in}^2}\right)(1.33)(0.9)(2.8 \text{ in})}$$
$$= 0.051 \text{ in}^2/\text{ft}$$

The problem statement gives the bar spacing as 32 in. A no. 4 bar has an area of 0.20 in^2. Try using no. 4 bars at 32 in.

$$A_{s,\text{prov}} = \left(\frac{0.20 \text{ in}^2}{32 \text{ in}}\right)\left(12 \; \frac{\text{in}}{\text{ft}}\right) = 0.075 \text{ in}^2/\text{ft}$$

MSJC code Sec. 2.3.3.3 limits the effective compressive width per bar to the lesser of

(a) center-to-center bar spacing: 32 in

(b) six times the wall thickness: (6)(6 in) = 36 in

(c) 72 in

Therefore, the effective compressive width per bar is the center-to-center bar spacing, or 32 in. Calculate the stresses based on this effective width.

To check the masonry stresses, first calculate j, k, and ρ. For a 32 in length of wall,

$$A_s = 0.20 \text{ in}^2$$
$$\rho = \frac{A_s}{bd} = \frac{0.20 \text{ in}^2}{(32 \text{ in})(2.8 \text{ in})}$$
$$= 0.00223$$
$$n = \frac{E_s}{E_m} = 21.5 \quad \text{[given]}$$
$$k = \sqrt{2\rho n + (\rho n)^2} - \rho n$$
$$= \sqrt{(2)(0.00223)(21.5) + \left((0.00223)(21.5)\right)^2}$$
$$\quad - (0.00223)(21.5)$$
$$= 0.265$$
$$j = 1 - \frac{k}{3} = 1 - \frac{0.265}{3}$$
$$= 0.912$$

When the neutral axis falls within the face shell, kd is less than the face-shell thickness and the analysis is the same as for a fully grouted masonry wall. See the *Civil Engineering Reference Manual* for further discussion.

$$kd = (0.265)(2.8 \text{ in})$$
$$= 0.74 \text{ in} < 1.0 \text{ in face shell} \quad \text{[OK]}$$

Calculate the allowable moment based on flexural compressive stress in the masonry for a 32 in length of wall. Take the $\frac{1}{3}$ allowable stress increase for wind.

$$F_b = \tfrac{1}{3}f'_m(1.33)$$
$$= \left(\frac{1}{3}\right)\left(1500 \; \frac{\text{lbf}}{\text{in}^2}\right)(1.33)$$
$$= 665 \text{ lbf/in}^2$$
$$M_m = \tfrac{1}{2}F_b bd^2 jk$$
$$= \left(\frac{1}{2}\right)\left(665 \; \frac{\text{lbf}}{\text{in}^2}\right)(32 \text{ in})(2.8 \text{ in})^2$$
$$\quad \times (0.912)(0.265)$$
$$= 20{,}160 \text{ in-lbf} \quad (20{,}000 \text{ in-lbf})$$
$$M_{\max} = \left(4092 \; \frac{\text{in-lbf}}{\text{ft}}\right)(32 \text{ in})\left(\frac{1 \text{ ft}}{12 \text{ in}}\right)$$
$$= 10{,}912 \text{ in-lbf}$$
$$M_m > M_{\max} \quad \text{[OK]}$$

Check the allowable moment based on the flexural tensile stress in the steel. Take the $\frac{1}{3}$ allowable stress increase for wind.

$$F_s = \left(24{,}000 \; \frac{\text{lbf}}{\text{in}^2}\right)(1.33)$$
$$= 31{,}920 \text{ lbf/in}^2$$

$$M_s = A_s F_s j d$$

$$= \left(0.075 \; \frac{\text{in}^2}{\text{ft}}\right)\left(31{,}920 \; \frac{\text{lbf}}{\text{in}^2}\right)(0.912)(2.8 \; \text{in})$$

$$= 6113 \; \text{in-lbf/ft} > 4092 \; \text{in-lbf/ft} = M_{\max} \quad [\text{OK}]$$

The allowable moment per foot based on the flexural compressive stress in the masonry is $M_m = 20160 \text{ in-lbf}$

$\times \; 12 \; \text{in}/\text{ft} \times \frac{1}{32} \; \text{in}$

$= 7560 \; \text{in-lbf/ft}$

(7600 in-lbf/ft)

The answer is (C).

Why Other Options Are Wrong

(A) This incorrect solution does not account for the $1/3$ allowable stress increase for wind when determining the amount of reinforcement required and checking the stresses in the masonry.

(B) This incorrect solution gives the answer for the allowable moment based on the tensile stress in the steel (6100 in-lbf/ft) instead of the allowable compressive flexural moment (7600 in-lbf/ft).

(D) This incorrect solution does not account for the spacing of the reinforcement when determining the area of steel provided per foot. This solution also fails to check the location of the neutral axis to confirm that the wall can be treated as a fully grouted wall.

SOLUTION 74

The beam-to-post connection shown imparts a lateral load on the wood screws. Because two wood members are joined in this connection, the connection is subjected to single shear. Appendix I of the NDS illustrates connections in single and double shear. Use NDS Chs. 10 and 11 and NDS Table 11L to determine the allowable lateral load on the connection.

The allowable lateral design value for a single connector is given in NDS Table 10.3.1 as

$$Z' = Z C_D C_M C_t C_g C_\Delta C_{\text{eg}} C_{di} C_{tn}$$

In this case, C_D is 1.0 for dead load plus live load, since the load duration factor for the shortest duration load applies (NDS Sec. 2.3.2 and App. B). C_M is 0.7, since the moisture content of this deck built in a humid climate will exceed 19% (NDS Table 10.3.3). Values for C_g and C_Δ depend on the diameter of the screw, D. Since $D < 0.25$ in for a 12-gage screw ($D = 0.216$ in) C_g and C_Δ equal 1.0 (NDS Sec. 10.3.6.1 and 11.5.1). C_{eg}, C_{di}, and C_{tn} do not apply.

Therefore,

$$Z' = Z C_D C_M C_g C_\Delta = Z(1.0)(0.7)(1.0)(1.0)$$

The cut-thread wood screw design values are given in NDS Table 11L. First, determine the side-member thickness. The side member is the 2×8 beam. Using Table

1B of the NDS Supplement, determine that the side-member thickness is 1.5 in. Next, find the column for southern pine and the row for a 12-gage wood screw. The tabulated design value is

$$Z = 160 \; \text{lbf}$$

$$Z' = (160 \; \text{lbf})(1.0)(0.7)(1.0)(1.0)$$

$$= 112 \; \text{lbf} \quad [\text{per screw}]$$

The capacity of the connection is

$$Z'_{\text{total}} = (5 \; \text{screws})\left(112 \; \frac{\text{lbf}}{\text{screw}}\right)$$

$$= 560 \; \text{lbf}$$

The answer is (C).

Why Other Options Are Wrong

(A) This incorrect solution misreads NDS Table 11L and uses the value for a 10-gage screw ($Z = 127$ lbf) instead of a 12-gage screw.

(B) This incorrect solution uses the wrong load duration factor. The load duration factor for the shortest-duration load applies. This solution uses the smallest load duration factor, $C_D = 0.9$.

(D) This incorrect solution fails to apply the adjustment factors to the tabulated design value to determine the allowable design value.

SOLUTION 75

The equivalent axial load procedure found in Part 3 of the AISC ASD manual can be used to design beam-columns.

$$P_e = P + M_x m + M_y m U$$

$$M_x = P e_x = (400 \; \text{kips})(12 \; \text{in})\left(\frac{1 \; \text{ft}}{12 \; \text{in}}\right)$$

$$= 400 \; \text{ft-kips}$$

$$M_y = 0 \; \text{ft-kips}$$

From AISC ASD Table B, Part 3 (Column Design for Grade 50 Steel and effective lengths over 22 ft), determine

$$m = 1.7$$

$$P_e = 400 \; \text{kips} + (400 \; \text{ft-kips})\left(1.7 \; \frac{1}{\text{ft}}\right) + 0 \; \text{kips}$$

$$= 1080 \; \text{kips}$$

Enter the column tables with an unbraced length of 25 ft, Grade 50 steel, and try a W14 × 193 section. By interpolation, a W14 × 193 section has a capacity of 1145 kips at an unbraced length of 25 ft. Since m is 1.7 for a W14 section, no further iterations are needed.

The answer is (A).

Why Other Options Are Wrong

(B) This incorrect solution uses the wrong grade of steel—Grade 36 instead of Grade 50.

(C) This incorrect solution uses the wrong grade of steel and does not perform the necessary number of iterations. The procedure must be repeated until the value for m used in the equation for P_e matches the value for m for the final section selected.

(D) This incorrect solution assumes the eccentricity about the wrong axis.

SOLUTION 76

The load on the stringers is

$$w = D + L$$

The weight of the slab is

$$D = t\gamma = (4 \text{ in}) \left(\frac{1 \text{ ft}}{12 \text{ in}} \right) \left(150 \, \frac{\text{lbf}}{\text{ft}^3} \right)$$
$$= 50 \text{ lbf/ft}^2$$

The live load during construction as given by the model codes (BOCA, SBC, etc.) is

$$L = 50 \text{ lbf/ft}^2$$
$$w = D + L = 50 \, \frac{\text{lbf}}{\text{ft}^2} + 50 \, \frac{\text{lbf}}{\text{ft}^2}$$
$$= 100 \text{ lbf/ft}^2$$

The maximum spacing is the lesser of the spacing limited by the bending of the joist, s_j, the spacing limited by the bending of the stringer, s_s, or the spacing limited by the load to the post, s_p.

$$s_j = \sqrt{\frac{4466 \text{ lbf}}{w}} = \sqrt{\frac{4466 \text{ lbf}}{100 \, \frac{\text{lbf}}{\text{ft}^2}}}$$
$$= 6.68 \text{ ft}$$

$$s_s = \frac{523.4 \, \frac{\text{lbf}}{\text{ft}}}{w} = \frac{523.4 \, \frac{\text{lbf}}{\text{ft}}}{100 \, \frac{\text{lbf}}{\text{ft}^2}}$$
$$= 5.23 \text{ ft}$$

$$s_p = \frac{512.8 \, \frac{\text{lbf}}{\text{ft}}}{w} = \frac{512.8 \, \frac{\text{lbf}}{\text{ft}}}{100 \, \frac{\text{lbf}}{\text{ft}^2}}$$
$$= 5.13 \text{ ft} \quad (5.1 \text{ ft})$$

The maximum spacing of the stringer is 5.1 ft.

The answer is (A).

Why Other Options Are Wrong

(B) This incorrect solution does not properly convert the slab thickness to feet. A slab thickness of 0.25 ft is used instead of 0.33 ft in calculating the weight of the slab.

(C) This incorrect solution is the largest of the spacings rather than the smallest.

(D) This incorrect solution does not include the live load in the calculation of the load on the stringers, w.

SOLUTION 77

Statement I is true. Mat foundations are useful in areas where the basement is below the ground water table (GWT) because the mat foundation provides a water barrier, since it is a continuous concrete slab.

Statement II is true. Mat foundations require both top and bottom reinforcement because both positive and negative moments are developed in the foundation.

The answer is (B).

(A) This incorrect solution only identifies statement I as true and misses the fact that statement II is also true.

(C) This incorrect solution mistakenly identifies the two solutions that are false rather than those that are true. Statements III and IV are false. Mat foundations are suitable where settlements may be a problem or where settlements may be large due, in part, to their rigidity. The rigidity of the mat tends to bridge over areas of erratic soil types or voids.

(D) This incorrect solution wrongly identifies statements III and IV as true. Mat foundations are suitable where settlements may be a problem or where settlements may be large due, in part, to their rigidity. The rigidity of the mat tends to bridge over areas of variable soil types or voids. Statements III and IV are false.

SOLUTION 78

From a foundation design reference, the minimum depth of penetration, D_{min}, for the sheet piling shown is given by the equation

$$D_{min}{}^4 - \left(\frac{8H}{\gamma(k_p - k_a)b} \right) D_{min}{}^2 - \left(\frac{12HL}{\gamma(k_p - k_a)b} \right)$$
$$\times D_{min} - \left(\frac{2H}{\gamma(k_p - k_a)b} \right)^2 = 0 \text{ ft}^4$$

$$k_a = \tan^2 \left(45 - \frac{\phi}{2} \right) = \tan^2 \left(45 - \frac{30}{2} \right)$$
$$= 0.333$$

$$k_p = \tan^2\left(45 + \frac{\phi}{2}\right) = \tan^2\left(45 + \frac{30}{2}\right)$$
$$= 3.00$$
$$\gamma = 110 \text{ lbf/ft}^3$$

From the problem statement, the width of the sheet piling, b, is 2 ft; the single concentrated load, H, is 10 kips (10,000 lbf); and the length of the sheet piling above grade, L, is 10 ft.

$$D_{\min}{}^4 - \left(\frac{(8)(10{,}000 \text{ lbf})}{\left(110 \frac{\text{lbf}}{\text{ft}^3}\right)(3.00 - 0.333)(2 \text{ ft})}\right) D_{\min}{}^2$$
$$- \left(\frac{(12)(10{,}000 \text{ lbf})(10 \text{ ft})}{\left(110 \frac{\text{lbf}}{\text{ft}^3}\right)(3.00 - 0.333)(2 \text{ ft})}\right) D_{\min}$$
$$- \left(\frac{(2)(10{,}000 \text{ lbf})}{\left(110 \frac{\text{lbf}}{\text{ft}^3}\right)(3.00 - 0.333)(2 \text{ ft})}\right)^2$$
$$= 0 \text{ ft}^4$$

$$D_{\min}{}^4 - (136.3 \text{ ft}^2)D_{\min}{}^2 - (2045.2 \text{ ft}^3)D_{\min}$$
$$- 1161.9 \text{ ft}^4 = 0 \text{ ft}^4$$

This equation can be solved iteratively.

Try $D_{\min} = 10$ ft.

$$(10 \text{ ft})^4 - (136.3 \text{ ft}^2)(10 \text{ ft})^2 - (2045.2 \text{ ft}^3)(10 \text{ ft})$$
$$- 1161.9 \text{ ft}^4 = 0 \text{ ft}^4$$
$$- 25{,}244 \text{ ft}^4 \neq 0 \text{ ft}^4$$

Try $D_{\min} = 15$ ft.

$$(15 \text{ ft})^4 - (136.3 \text{ ft}^2)(15 \text{ ft})^2 - (2045.2 \text{ ft}^3)(15 \text{ ft})$$
$$- 1161.9 \text{ ft}^4 = 0 \text{ ft}^4$$
$$- 11{,}882 \text{ ft}^4 \neq 0 \text{ ft}^4$$

Try $D_{\min} = 16$ ft.

$$(16 \text{ ft})^4 - (136.3 \text{ ft}^2)(16 \text{ ft})^2 - (2045.2 \text{ ft}^3)(16 \text{ ft})$$
$$- 1161.9 \text{ ft}^4 = 0 \text{ ft}^4$$
$$- 3242 \text{ ft}^4 \neq 0 \text{ ft}^4$$

Try $D_{\min} = 16.3$ ft.

$$(16.3 \text{ ft})^4 - (136.3 \text{ ft}^2)(16.3 \text{ ft})^2 - (2045.2 \text{ ft}^3)(16.3 \text{ ft})$$
$$- 1161.9 \text{ ft}^4 = 0 \text{ ft}^4$$
$$- 121 \text{ ft}^4 \approx 0 \text{ ft}^4$$

This last solution is close to equaling zero and is precise enough for this example. Further iterations would yield an exact solution.

The minimum depth of penetration is

$$D_{\min} = 16.3 \text{ ft}$$

Applying the factor of safety of 1.3 yields

$$D = (1.3)(16.3 \text{ ft}) = 21.2 \text{ ft}$$

The total length of the sheet piling is

$$L_{\text{total}} = L + D = 10 \text{ ft} + 21.2 \text{ ft}$$
$$= 31.2 \text{ ft} \quad (31 \text{ ft})$$

The answer is (C).

Why Other Options Are Wrong

(A) This incorrect solution calculates the depth of penetration only, not the total length of the pile.

(B) This incorrect solution does not apply the factor of safety when calculating the required depth.

(D) This incorrect solution does not include the width of the sheet piling, b, in the equation for calculating the minimum depth of penetration, $D_{\min}$.

SOLUTION 79

The allowable bearing capacity of a single pile is

$$Q_a = \frac{Q_u}{F}$$

The ultimate bearing capacity is the sum of the point-bearing capacity, Q_p, and the skin-friction capacity, Q_f.

$$Q_u = Q_p + Q_f$$
$$Q_p = A_p c N_c$$

From a foundation handbook, for driven piles of virtually all conventional dimensions,

$$N_c = 9$$
$$Q_p = 9 A_p c$$
$$Q_f = A_s f_s$$
$$f_s = c_A + \sigma_h \tan \delta$$

For saturated clay, the angle of external friction, δ, is $0°$.

$$Q_f = A_s f_s = A_s c_A$$

The area of the pile tip for a steel H-pile is calculated using the area enclosed by the block perimeter. From the AISC ASD manual, Part I, find the block dimensions of an HP12 × 63 to be 11.94 in by 12.125 in.

$$A_p = db_f = \frac{(11.94 \text{ in})(12.125 \text{ in})}{144 \frac{\text{in}^2}{\text{ft}^2}}$$

$$= 1.01 \text{ ft}^2$$

$$Q_p = 9A_p c = \frac{(9)(1.01 \text{ ft}^2)\left(400 \frac{\text{lbf}}{\text{ft}^2}\right)}{1000 \frac{\text{lbf}}{\text{kip}}}$$

$$= 3.6 \text{ kips}$$

In an H-pile, the area between the flanges is assumed to fill with soil that moves with the pile. In calculating the skin area, A_s, of an H-pile, the perimeter is the block perimeter of the pile.

$$A_s = pL$$
$$= \big((2)(11.94 \text{ in}) + (2)(12.125 \text{ in})\big)$$
$$\times \left(\frac{1 \text{ ft}}{12 \text{ in}}\right)(100 \text{ ft})$$
$$= 401 \text{ ft}^2$$

$$Q_f = A_s c_A = \frac{(401 \text{ ft}^2)\left(360 \frac{\text{lbf}}{\text{ft}^2}\right)}{1000 \frac{\text{lbf}}{\text{kip}}}$$

$$= 144.4 \text{ kips}$$
$$Q_u = Q_p + Q_f = 3.6 \text{ kips} + 144.4 \text{ kips}$$
$$= 148.0 \text{ kips}$$

With a factor of safety of 3, the allowable load is

$$Q_a = \frac{Q_u}{F} = \frac{148.0 \text{ kips}}{3}$$
$$= 49.3 \text{ kips} \quad (49 \text{ kips})$$

The answer is (B).

Why Other Options Are Wrong

(A) This incorrect solution does not include the length of the pile in calculating the skin area. The units do not work out.

(C) This solution calculates the perimeter area of the pile incorrectly. When determining the pile capacity, the skin area and the point area of an H-pile should be calculated using the block perimeter and block area of the pile. In this wrong solution, the actual pile perimeter is used, as is the actual area of steel section.

(D) This incorrect solution does not include the factor of safety in determining the allowable capacity.

SOLUTION 80

Chapter 6 of the NDS contains the specifications for round timber piles. The allowable compression design value is

$$F'_c = F_c C_D C_t C_u C_p C_{\text{cs}} C_{\text{sp}}$$

From NDS Table 2.3.3, for in-service temperatures less than 100°F,

$$C_t = 1.0$$

From NDS Table 6.3.5, for kiln dried piles,

$$C_u = 1.11$$

From NDS Table 6.3.12, for pile clusters,

$$C_{\text{sp}} = 1.0$$
$$C_p = 0.62 \quad \text{[given]}$$

From NDS Sec. 6.3.10, for round timber piles made from red oak,

$$C_{\text{cs}} = 1.0$$
$$F_c = 1100 \text{ lbf/in}^2 \quad \text{[NDS Table 6A]}$$

The load duration factor depends upon the type of load. The load duration factor for the shortest duration load in a combination of loads shall apply for that load combination (NDS Sec. 2.3.2.2). There are two load cases to consider: dead load, and dead load plus live load.

Dead Load Only:

$$C_D = 0.9$$
$$F'_c = F_c C_D C_t C_u C_p C_{\text{cs}} C_{\text{sp}}$$
$$= \left(1100 \frac{\text{lbf}}{\text{in}^2}\right)(0.9)(1.0)(1.11)(0.62)(1.0)(1.0)$$
$$= \left(1100 \frac{\text{lbf}}{\text{in}^2}\right)(0.619)$$
$$= 681 \text{ lbf/in}^2$$

Dead Load plus Live Load:

$$C_D = 1.0$$
$$F'_c = F_c C_D C_t C_u C_p C_{\text{cs}} C_{\text{sp}}$$
$$= \left(1100 \frac{\text{lbf}}{\text{in}^2}\right)(1.0)(1.0)(1.11)(0.62)(1.0)(1.0)$$
$$= \left(1100 \frac{\text{lbf}}{\text{in}^2}\right)(0.688)$$
$$= 757 \text{ lbf/in}^2$$

Determine the critical load case.

Dead Load Only:

$$f_{c,D} = \frac{P_D}{A} = \frac{\left(\dfrac{300 \text{ kips}}{3 \text{ piles}}\right)\left(1000 \dfrac{\text{lbf}}{\text{kip}}\right)}{230 \text{ in}^2}$$
$$= 435 \text{ lbf/in}^2 \text{ per pile}$$
$$f_{c,D} < F_c' \quad [\text{OK}]$$

Dead Load plus Live Load:

$$f_{c,D+L} = \frac{P_{D+L}}{A}$$
$$= \frac{\left(\dfrac{300 \text{ kips} + 400 \text{ kips}}{3 \text{ piles}}\right)\left(1000 \dfrac{\text{lbf}}{\text{kip}}\right)}{230 \text{ in}^2}$$
$$= 1014 \text{ lbf/in}^2 \text{ per pile}$$
$$f_{c,D+L} > F_c' \quad [\text{no good}]$$

The critical load case is dead load plus live load. The total adjustment factor for this load case is 0.688.

The answer is (D).

Why Other Options Are Wrong

(A) This incorrect solution includes the single pile factor (NDS Sec. 6.3.12). Although the stress is calculated for a single pile, the pile is still one in a group, and this factor does not apply.

(B) This incorrect solution uses the *smallest* load duration factor (0.9) for the dead plus live load combination instead of using the load duration factor for the *shortest* load (1.0).

(C) This incorrect solution does not include the untreated factor, C_u.

SOLUTION 81

Begin by determining the lateral earth pressure on the wall using the Rankine active pressure, k_a.

The lateral earth pressure due to the surcharge from the concrete slab is

$$p_q = k_a q$$
$$= (0.361)\left((10 \text{ in})\left(\frac{1 \text{ ft}}{12 \text{ in}}\right)\left(0.150 \frac{\text{kip}}{\text{ft}^3}\right)\right)$$
$$= 0.045 \text{ kip/ft}^2$$

The lateral earth pressure due to the soil backfill is

$$p_a = k_a p_v = k_a \gamma H$$

At the bottom of the wall,

$$H = 20 \text{ ft}$$
$$p_a = k_a \gamma H = (0.361)\left(0.110 \frac{\text{kip}}{\text{ft}^3}\right)(20 \text{ ft})$$
$$= 0.794 \text{ kip/ft}^2$$

The combined lateral pressure is

$$p_{a,\text{total}} = k_a q + k_a \gamma H = 0.045 \frac{\text{kip}}{\text{ft}^2}$$
$$+ \left(0.0397 \frac{\text{kip}}{\text{ft}^3}\right) H$$

At the top of the wall,

$$p_{a,\text{total}} = 0.045 \frac{\text{kip}}{\text{ft}^2} + \left(0.0397 \frac{\text{kip}}{\text{ft}^3}\right)(0 \text{ ft})$$
$$= 0.045 \text{ kip/ft}^2$$

At the bottom of the wall,

$$p_{a,\text{total}} = 0.045 \frac{\text{kip}}{\text{ft}^2} + \left(0.0397 \frac{\text{kip}}{\text{ft}^3}\right)(20 \text{ ft})$$
$$= 0.839 \text{ kip/ft}^2$$

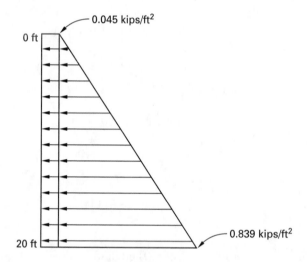

The total active resultant is

$$R_a = k_a q H + \tfrac{1}{2} k_a \gamma H^2$$
$$= \left(0.045 \frac{\text{kip}}{\text{ft}^2}\right)(20 \text{ ft}) + \left(\frac{1}{2}\right)$$
$$\times \left(0.0397 \frac{\text{kip}}{\text{ft}^3}\right)(20 \text{ ft})^2$$
$$= 8.84 \text{ kips/ft}^2$$

The horizontal resultant is located at a distance y_a from the toe.

$$y_a = \frac{\sum R_a y}{\sum R_a}$$
$$= \frac{\left(0.90 \dfrac{\text{kip}}{\text{ft}}\right)\left(\dfrac{20 \text{ ft}}{2}\right) + \left(7.94 \dfrac{\text{kips}}{\text{ft}}\right)\left(\dfrac{20 \text{ ft}}{3}\right)}{0.90 \dfrac{\text{kips}}{\text{ft}} + 7.94 \dfrac{\text{kips}}{\text{ft}}}$$
$$= 7.01 \text{ ft}$$

Calculate the overturning moment about the toe.

$$M_{\text{overturning}} = R_a y_a$$
$$= \left(8.84 \ \frac{\text{kips}}{\text{ft}^2}\right)(7.01 \text{ ft})$$
$$= 62.0 \text{ ft-kips/ft}$$

Overturning is resisted by the weight of the soil acting vertically on the footing, the weight of the retaining wall, and the surcharge. Passive restraint from the soil in front of the footing is only considered if it will always be there. In most cases it is neglected.

$$M_{\text{resisting}} = \sum W_i x_i$$
$$W_i = \gamma_i A_i$$

Soil:

$$A_{\text{soil}} = LH$$
$$= \left(15 \text{ ft} - (20 \text{ in})\left(\frac{1 \text{ ft}}{12 \text{ in}}\right)\right)(20 \text{ ft} - 2 \text{ ft})$$
$$= 240 \text{ ft}^2$$
$$W_{\text{soil}} = \gamma_{\text{soil}} A_{\text{soil}}$$
$$= \left(0.110 \ \frac{\text{kip}}{\text{ft}^3}\right)(240 \text{ ft}^2)$$
$$= 26.4 \text{ kips/ft}$$

Retaining Wall, Vertical Section:

$$A_{w,v} = LH$$
$$= (20 \text{ in})\left(\frac{1 \text{ ft}}{12 \text{ in}}\right)(20 \text{ ft})$$
$$= 33.3 \text{ ft}^2$$
$$W_{w,v} = \gamma_w A_w$$
$$= \left(0.150 \ \frac{\text{kip}}{\text{ft}^3}\right)(33.3 \text{ ft}^2)$$
$$= 5.00 \text{ kips/ft}$$

Retaining Wall, Horizontal Section:

$$A_{w,h} = LH$$
$$= (13.33 \text{ ft})(2 \text{ ft})$$
$$= 26.7 \text{ ft}^2$$
$$W_{w,h} = \gamma_w A_w$$
$$= \left(0.150 \ \frac{\text{kip}}{\text{ft}^3}\right)(26.7 \text{ ft}^2)$$
$$= 4.01 \text{ kips/ft}$$

Concrete Slab Surcharge:

$$A_{\text{slab}} = LH$$
$$= (8 \text{ ft})(10 \text{ in})\left(\frac{1 \text{ ft}}{12 \text{ in}}\right)$$
$$= 6.67 \text{ ft}^2$$
$$W_{\text{slab}} = \gamma_{\text{slab}} A_{\text{slab}}$$
$$= \left(0.150 \ \frac{\text{kip}}{\text{ft}^3}\right)(6.67 \text{ ft}^2)$$
$$= 1.00 \text{ kip/ft}$$

The resisting moment about the toe is

$$M_{\text{resisting}} = \sum W_i x_i$$
$$= W_{\text{soil}} x_{\text{soil}} + W_{w,v} x_{w,v}$$
$$\qquad + W_{w,h} x_{w,h} + W_{\text{slab}} x_{\text{slab}}$$
$$= \left(26.4 \ \frac{\text{kips}}{\text{ft}}\right)\left(\frac{13.33 \text{ ft}}{2} + (20 \text{ in})\left(\frac{1 \text{ ft}}{12 \text{ in}}\right)\right)$$
$$\quad + \left(5.00 \ \frac{\text{kips}}{\text{ft}}\right)\left(\frac{1}{2}\right)(20 \text{ in})\left(\frac{1 \text{ ft}}{12 \text{ in}}\right)$$
$$\quad + \left(4.01 \ \frac{\text{kips}}{\text{ft}}\right)\left(\begin{array}{c}\dfrac{13.33 \text{ ft}}{2} + (20 \text{ in}) \\ \times \left(\dfrac{1 \text{ ft}}{12 \text{ in}}\right)\end{array}\right)$$
$$\quad + \left(1.00 \ \frac{\text{kip}}{\text{ft}}\right)\left(\frac{8 \text{ ft}}{2} + (20 \text{ in})\left(\frac{1 \text{ ft}}{12 \text{ in}}\right)\right)$$
$$= 263 \text{ ft-kips/ft}$$

The factor of safety against overturning is

$$F_{\text{OT}} = \frac{M_{\text{resisting}}}{M_{\text{overturning}}} = \frac{263 \ \frac{\text{ft-kips}}{\text{ft}}}{62.0 \ \frac{\text{ft-kips}}{\text{ft}}}$$
$$= 4.24$$

The answer is (B).

Why Other Options Are Wrong

(A) This incorrect solution correctly calculates the overturning moment but neglects the effect of the slab in calculating the resisting moment.

(C) This incorrect solution miscalculates the area of the soil when determining the resisting moment by using the full wall height instead of the soil height.

(D) This incorrect solution does not include the effects of the surcharge in determining the overturning moment or the resisting moment.

SOLUTION 82

The sliding force is resisted by friction and adhesion between the soil and the base. Passive restraint from the soil is only considered if it will always be there. In most cases it is neglected. The sliding force is the sum of the pressure due to the earth and the pressure due to the slab surcharge. From the *Civil Engineering Reference Manual* or another reference book,

$$R_{a,h} = \tfrac{1}{2}k_a\gamma H^2 + k_a q H$$

The lateral earth pressure due to the soil backfill is

$$
\begin{aligned}
R_{a,h,\text{soil}} &= \tfrac{1}{2}k_a\gamma H^2 \\
&= \left(\frac{1}{2}\right)(0.361)\left(0.110\ \frac{\text{kip}}{\text{ft}^3}\right)(20\ \text{ft})^2 \\
&= 7.94\ \text{kips/ft}
\end{aligned}
$$

The lateral earth pressure due to the surcharge from the concrete slab is

$$
\begin{aligned}
R_{a,h,\text{surcharge}} &= k_a q H = (0.361)\left((10\ \text{in})\left(\frac{1\ \text{ft}}{12\ \text{in}}\right)\right. \\
&\quad \left.\times \left(0.150\ \frac{\text{kip}}{\text{ft}^3}\right)\right)(20\ \text{ft}) \\
&= 0.90\ \text{kip/ft}
\end{aligned}
$$

$$
\begin{aligned}
R_{a,h} &= \tfrac{1}{2}k_a\gamma H^2 + k_a q H \\
&= 7.94\ \frac{\text{kips}}{\text{ft}} + 0.90\ \frac{\text{kip}}{\text{ft}} \\
&= 8.84\ \text{kips/ft}
\end{aligned}
$$

Sliding is resisted by friction from the weight of the soil, retaining wall, and surcharge. From a reference handbook such as the *Civil Engineering Reference Manual*,

$$
\begin{aligned}
R_{\text{SL}} &= \left(\sum W_i + R_{a,v}\right)\tan\delta + c_A B \\
W_i &= \gamma_i A_i
\end{aligned}
$$

Since the backfill has no slope, there is no vertical component of the active earth pressure.

$$R_{a,v} = 0\ \text{kips/ft}$$

According to a reference handbook, without other information, $\tan\delta$ is approximately 0.55 for sandy soil without silt.

$$
\begin{aligned}
c_A &= 0\ \text{kips/ft}^2 \quad \text{[for granular soil]} \\
R_{\text{SL}} &= \left(\sum W_i\right)\tan\delta \\
&= \left(\sum W_i\right)(0.55)
\end{aligned}
$$

Determine the weight of the soil, wall, and surcharge.

Soil:

$$
\begin{aligned}
A_{\text{soil}} &= LH = \left(15\ \text{ft} - (20\ \text{in})\left(\frac{1\ \text{ft}}{12\ \text{in}}\right)\right) \\
&\quad \times (20\ \text{ft} - 2\ \text{ft}) \\
&= 240\ \text{ft}^2 \\
W_{\text{soil}} &= \gamma_{\text{soil}} A_{\text{soil}} \\
&= \left(0.110\ \frac{\text{kip}}{\text{ft}^3}\right)(240\ \text{ft}^2) \\
&= 26.4\ \text{kips/ft}
\end{aligned}
$$

Retaining Wall, Vertical Section:

$$
\begin{aligned}
A_{w,v} &= LH = (20\ \text{in})\left(\frac{1\ \text{ft}}{12\ \text{in}}\right)(20\ \text{ft}) \\
&= 33.3\ \text{ft}^2 \\
W_{w,v} &= \gamma_w A_w = \left(0.150\ \frac{\text{kip}}{\text{ft}^3}\right)(33.3\ \text{ft}^2) \\
&= 5.00\ \text{kips/ft}
\end{aligned}
$$

Retaining Wall, Horizontal Section:

$$
\begin{aligned}
A_{w,h} &= LH \\
&= (13.33\ \text{ft})(2\ \text{ft}) \\
&= 26.7\ \text{ft}^2 \\
W_{w,h} &= \gamma_w A_w \\
&= \left(0.150\ \frac{\text{kip}}{\text{ft}^3}\right)(26.7\ \text{ft}^2) \\
&= 4.01\ \text{kips/ft}
\end{aligned}
$$

Concrete Slab Surcharge:

$$
\begin{aligned}
A_{\text{slab}} &= LH \\
&= (8\ \text{ft})(10\ \text{in})\left(\frac{1\ \text{ft}}{12\ \text{in}}\right) \\
&= 6.67\ \text{ft}^2 \\
W_{\text{slab}} &= \gamma_{\text{slab}} A_{\text{slab}} \\
&= \left(0.150\ \frac{\text{kip}}{\text{ft}^3}\right)(6.67\ \text{ft}^2) \\
&= 1.00\ \text{kip/ft}
\end{aligned}
$$

$$
\begin{aligned}
\sum W_i &= W_{\text{soil}} + W_{w,v} + W_{w,h} + W_{\text{slab}} \\
&= 26.4\ \frac{\text{kips}}{\text{ft}} + 5.00\ \frac{\text{kips}}{\text{ft}} \\
&\quad + 4.01\ \frac{\text{kips}}{\text{ft}} + 1.00\ \frac{\text{kip}}{\text{ft}} \\
&= 36.4\ \text{kips/ft}
\end{aligned}
$$

The resistance against sliding is

$$
\begin{aligned}
R_{\text{SL}} &= \left(\sum W_i\right)\tan\delta = \left(36.4\ \frac{\text{kips}}{\text{ft}}\right)(0.55) \\
&= 20.0\ \text{kips/ft}
\end{aligned}
$$

The factor of safety against sliding is

$$F_{\rm SL} = \frac{R_{\rm SL}}{R_{a,h}} = \frac{20.0 \ \frac{\rm kips}{\rm ft}}{8.84 \ \frac{\rm kips}{\rm ft}}$$

$$= 2.26$$

The answer is (B).

Why Other Options Are Wrong

(A) In this incorrect solution, the weight of the slab is not included when calculating the resisting force.

(C) This incorrect solution neglects the effect of the surcharge entirely.

(D) This incorrect solution includes the restraint provided by the passive force. Unless explicitly stated that it will always be there, passive restraint should be ignored.

SOLUTION 83

Section 11.10 of the ACI 318 covers design of shear walls. Since the wall supports a uniform load, w_u, that is positive, the wall is in compression and the concrete shear strength is

$$V_c = 2\sqrt{f'_c}hd \quad \text{[ACI 318 Sec. 11.10.5]}$$
$$V_c = 2\sqrt{f'_c}hd$$
$$= (2)\left(\sqrt{4000} \ \frac{\rm lbf}{\rm in^2}\right)(12 \ \rm in)(110 \ \rm in)\left(\frac{1 \ \rm kip}{1000 \ \rm lbf}\right)$$
$$= 167 \ \rm kips$$
$$V_n = V_c + V_s$$
$$= 167 \ \rm kips + 700 \ \rm kips$$
$$= 867 \ \rm kips$$

The nominal shear strength of the wall is limited by ACI 318 Sec. 11.10.3 to

$$V_n = 10\sqrt{f'_c}hd$$
$$= (10)\left(\sqrt{4000} \ \frac{\rm lbf}{\rm in^2}\right)(12 \ \rm in)(110 \ \rm in)\left(\frac{1 \ \rm kip}{1000 \ \rm lbf}\right)$$
$$= 834.8 \ \rm kips \quad \text{[controls]}$$

Therefore, the nominal shear strength of the wall is 834.8 kips (830 kips).

The answer is (C).

Why Other Options Are Wrong

(A) This incorrect solution calculates the shear strength of the concrete instead of the shear strength of the wall.

(B) This incorrect solution calculates the factored shear strength of the wall.

(D) This incorrect solution ignores the limit on nominal shear strength found in ACI 318 Sec. 11.10.3.

SOLUTION 84

From ACI 318 Sec. 11.10.8, the amount of reinforcement required depends upon whether or not $V_u < \phi V_c/2$.

$$\phi V_c = \phi 2\sqrt{f'_c}hd$$
$$\phi = \cancel{0.85} \ \ 0.75$$
$$h = 10 \ \rm in$$
$$d = 110 \ \rm in$$
$$\phi V_c = \phi 2\sqrt{f'_c}hd = (0.85)(2)\left(\sqrt{6000} \ \frac{\rm lbf}{\rm in^2}\right)(10 \ \rm in)$$
$$\times (110 \ \rm in)\left(\frac{1 \ \rm kip}{1000 \ \rm lbf}\right)$$
$$= \underset{128}{\cancel{145}} \ \rm kips$$
$$\frac{\phi V_c}{2} = \frac{145 \ \rm kips}{2}$$
$$= \underset{64}{\cancel{72.5}} \ \rm kips < V_u$$
$$= 110 \ \rm kips$$

Therefore, the reinforcement must be designed in accordance with ACI 318 Sec. 11.10.9. Since $\phi V_c > V_u$, the area of reinforcement is controlled by the minimum area requirements.

ACI 318 Sec. 11.10.9.2 states that the ratio of horizontal shear reinforcement area to the gross concrete area of the vertical section shall not be less than 0.0025.

$$\rho_h = \frac{A_v}{A_g} = 0.0025$$
$$A_v = 0.0025 A_g = 0.0025 h h_w$$
$$= (0.0025)(10 \ \rm in)(30 \ \rm ft)\left(12 \ \frac{\rm in}{\rm ft}\right)$$
$$= 9.00 \ \rm in^2$$
$$\frac{A_v}{s_2} = \frac{9.00 \ \rm in^2}{30 \ \rm ft}$$
$$= 0.30 \ \rm in^2/ft$$

Number 5 bars at 12 in spacing provide 0.31 in²/ft.

Check the spacing limits in ACI 318 Sec. 11.10.9.3. The maximum spacing is the lesser of

$$s_2 \le \frac{l_w}{5} = \frac{120 \ \rm in}{5} = 24 \ \rm in$$
$$s_2 \le 3h = (3)(10 \ \rm in) = 30 \ \rm in$$
$$s_2 \le 18 \ \rm in \quad \text{[controls]}$$

Number 5 at 12 in is OK.

The answer is (C).

Why Other Options Are Wrong

(A) This incorrect solution uses the horizontal cross section, instead of the vertical section, in calculating A_g.

(B) This incorrect solution calculates V_s incorrectly as

$$V_s = \phi V_c - V_u$$
$$= 145 \text{ kips} - 110 \text{ kips}$$
$$= 35 \text{ kips}$$

(D) This incorrect solution uses the minimum horizontal reinforcement requirements found in ACI 318 Ch. 14. ACI 318 Ch. 14 can only be used when $V_u < (\phi V_c)/2$.

SOLUTION 85

The design axial load strength of a wall is given in ACI 318 Sec. 14.5.2 as

$$\phi P_{\text{nw}} = 0.55 \phi f_c' A_g \left(1 - \left(\frac{k l_c}{32 h} \right)^2 \right) \quad \text{[ACI 318 Eq. 14-1]}$$

Calculate the factored axial load and set it equal to the design axial strength.

$$P_u = w_u l_w = \left(5 \frac{\text{kips}}{\text{ft}} \right) (16 \text{ ft}) = 80 \text{ kips}$$

$$P_u = \phi P_{\text{nw}}$$

$$= 0.55 \phi f_c' A_g \left(1 - \left(\frac{k l_c}{32 h} \right)^2 \right)$$

$$= 0.55 \phi f_c' (h l_w) \left(1 - \left(\frac{k l_c}{32 h} \right)^2 \right)$$

$$\phi = \underset{0.70}{\cancel{0.65}} \quad \text{[ACI 318 Sec. 9.3.2.2]}$$

$$80 \text{ kips} = (0.55)(0.65) \left(4000 \frac{\text{lbf}}{\text{in}^2} \right) \left(\frac{1 \text{ kip}}{1000 \text{ lbf}} \right)$$

$$\times (12 \text{ in})(16 \text{ ft}) \left(12 \frac{\text{in}}{\text{ft}} \right)$$

$$\times \left(1 - \left(\frac{(1.0) l_c}{(32)(12 \text{ in})} \right)^2 \right)$$

$$= (3295 \text{ kips}) \left(1 - \frac{l_c^2}{147{,}456 \text{ in}^2} \right)$$

$$0.024279 = 1 - \frac{l_c^2}{147{,}456 \text{ in}^2}$$

$$\frac{l_c^2}{147{,}456 \text{ in}^2} = 0.975721$$

$$l_c = \underset{379.6}{\cancel{(379.3} \text{ in})} \left(\frac{1 \text{ ft}}{12 \text{ in}} \right)$$

$$= 31.6 \text{ ft}$$

Check the minimum thickness requirements found in ACI 318 Sec. 14.5.3. Thickness is based on the shorter of the wall height and wall length. If the wall height is 31.6 ft and the wall length is 16 ft, the minimum thickness is

$$h > \frac{l_w}{25} = \frac{(16 \text{ ft}) \left(12 \frac{\text{in}}{\text{ft}} \right)}{25}$$

$$= 7.7 \text{ in} \quad \text{[OK]}$$

The answer is (C).

Why Other Options Are Wrong

(A) This incorrect solution does not properly apply the requirements for minimum thickness found in ACI 318 Sec. 14.5.3. The maximum height is incorrectly calculated as the product of 25 times the wall thickness, h. This requirement should be used to check the minimum thickness based upon the lesser of the height or length. In this case, the length, not the height, would be the limiting value.

(B) This incorrect solution calculates the gross area with the length of the wall in feet rather than inches. The units do not work out.

(D) This incorrect solution calculates the uniform axial load, w_u, rather than the total axial load, P_u. The units do not work out.

SOLUTION 86

The ratio of vertical shear reinforcement area to gross concrete area for a shear wall is given in ACI 318 Sec. 11.10.9.4 as

$$\rho_n = 0.0025 + 0.5 \left(2.5 - \frac{h_w}{l_w} \right) (\rho_h - 0.0025)$$

$$\text{[ACI 318 Eq. 11-32]}$$

This value need not exceed the required horizontal shear reinforcement, ρ_h.

$$h_w = (10 \text{ ft}) \left(12 \frac{\text{in}}{\text{ft}} \right) = 120 \text{ in}$$

$$l_w = (30 \text{ ft}) \left(12 \frac{\text{in}}{\text{ft}} \right) = 360 \text{ in}$$

$$\rho_n = 0.0025 + 0.5 \left(2.5 - \frac{h_w}{l_w} \right) (\rho_h - 0.0025)$$

$$= 0.0025 + (0.5) \left(2.5 - \frac{120 \text{ in}}{360 \text{ in}} \right) (0.0040 - 0.0025)$$

$$= 0.00413$$

$$\rho_h = 0.0040$$

Since ρ_n must be less than or equal to ρ_h, ρ_n is 0.00400.

The answer is (C).

Why Other Options Are Wrong

(A) This incorrect solution reverses the height and length of the wall in ACI 318 Eq. 11-32 and ignores the minimum reinforcement requirement of ACI 318 Sec. 11.10.9.4.

(B) This incorrect solution reverses the height and length of the wall in ACI 318 Eq. 11-32.

(D) This incorrect solution correctly calculates the vertical shear reinforcement ratio but does not apply the limit of $\rho_n \leq \rho_h$ given in ACI 318 Sec. 11.10.9.4.

SOLUTION 87

The minimum applied axial load is limited by the tensile stress in the masonry. Section 2.2.3.2 of the MSJC code allows flexural tension transverse to the plane of masonry only for unreinforced masonry. Therefore, no flexural tension is allowed in the plane of the unreinforced masonry. The resultant stress must be in compression.

$$f = \frac{P}{A} \pm \frac{M}{S} \quad \text{[must be in compression]}$$

The maximum applied axial load is limited by the compressive stress in the masonry.

$$\frac{f_a}{F_a} + \frac{f_b}{F_b} \leq 1.0 \quad \text{[MSJC Eq. 2-10]}$$

Calculate the section properties of the wall. Hollow masonry units are face-shell bedded only. Section properties are based on face-shell dimensions only. From the *Masonry Designers' Guide* (MDG-4) App. A or a similar reference, determine that the face-shell thickness of a 12 in unit is 1.5 in. For in-plane bending,

$$A_n = (1.5 \text{ in})(17.7 \text{ ft})\left(12 \frac{\text{in}}{\text{ft}}\right)(2)$$
$$= 637 \text{ in}^2$$

$$S_n = \frac{bd^2}{6} = \frac{(2)(1.5 \text{ in})\left((17.7 \text{ ft})\left(12 \frac{\text{in}}{\text{ft}}\right)\right)^2}{6}$$
$$= 22{,}557 \text{ in}^3$$

$$I_n = \frac{bd^3}{12} = \frac{(2)(1.5 \text{ in})\left((17.7 \text{ ft})\left(12 \frac{\text{in}}{\text{ft}}\right)\right)^3}{12}$$
$$= 2{,}395{,}540 \text{ in}^4$$

$$r = \sqrt{\frac{I_n}{A_n}} = \sqrt{\frac{2{,}395{,}540 \text{ in}^4}{637 \text{ in}^2}}$$
$$= 61.3 \text{ in}$$

$$\frac{h}{r} = \frac{(20.0 \text{ ft})\left(12 \frac{\text{in}}{\text{ft}}\right)}{61.3 \text{ in}}$$
$$= 3.9 < 99$$

$$M = Wh = (3200 \text{ lbf})(18 \text{ ft})$$
$$= 57{,}600 \text{ ft-lbf}$$

$$P_{\text{self-wt}} = whl = \left(51.0 \frac{\text{lbf}}{\text{ft}^2}\right)(20.0 \text{ ft})(17.7 \text{ ft})$$
$$= 18{,}054 \text{ lbf}$$

Calculate the minimum applied axial load. This load is limited by zero tension in the masonry.

$$-\frac{P}{A_n} + \frac{M}{S} = 0 \text{ lbf/in}^2$$

$$-\frac{P}{637 \text{ in}^2} + \frac{(57{,}600 \text{ ft-lbf})\left(12 \frac{\text{in}}{\text{ft}}\right)}{22{,}557 \text{ in}^3} = 0 \text{ lbf/in}^2$$

$$-\frac{P}{637 \text{ in}^2} = -30.64 \frac{\text{lbf}}{\text{in}^2}$$
$$P = 19{,}519 \text{ lbf}$$

$$P_{\text{applied}} = P - P_{\text{self-wt}}$$
$$= 19{,}519 \text{ lbf} - 18{,}054 \text{ lbf}$$
$$= 1465 \text{ lbf} \quad (1500 \text{ lbf})$$

Since the compressive stresses are so small, the interaction equation (MSJC Eq. 2-10) is OK by inspection.

Calculate the maximum applied load. This load is limited by compressive stress in the masonry.

$$\frac{f_a}{F_a} + \frac{f_b}{F_b} \leq 1.0 \quad \text{[MSJC Eq. 2-10]}$$

$$f_a = \frac{P}{A_n} = \frac{P}{637 \text{ in}^2}$$

$$F_a = \tfrac{1}{4} f'_m \left(1 - \left(\frac{h}{140r}\right)^2\right)$$
$$= \left(\frac{1}{4}\right)\left(2500 \frac{\text{lbf}}{\text{in}^2}\right)\left(1 - \left(\frac{3.9}{140}\right)^2\right)$$
$$= 625 \text{ lbf/in}^2 \quad \text{[MSJC Eq. 2-12]}$$

$$f_b = \frac{M}{S_n} = \frac{(57{,}600 \text{ ft-lbf})\left(12 \frac{\text{in}}{\text{ft}}\right)}{22{,}557 \text{ in}^3}$$
$$= 30.6 \text{ lbf/in}^2$$

$$F_b = \tfrac{1}{3}f'_m = \left(\tfrac{1}{3}\right)\left(2500 \ \frac{\text{lbf}}{\text{in}^2}\right)$$
$$= 833 \ \text{lbf/in}^2 \quad [\text{MSJC Eq. 2-14}]$$

$$\frac{f_a}{F_a} + \frac{f_b}{F_b} = \frac{\dfrac{P}{637 \ \text{in}^2}}{625 \ \dfrac{\text{lbf}}{\text{in}^2}} + \frac{30.6 \ \dfrac{\text{lbf}}{\text{in}^2}}{833 \ \dfrac{\text{lbf}}{\text{in}^2}} = 1.0$$

$$\frac{P}{637 \ \text{in}^2} = 602.0 \ \text{lbf/in}^2$$
$$P = 383{,}500 \ \text{lbf}$$
$$P_{\text{applied}} = P - P_{\text{self-wt}} = 383{,}500 \ \text{lbf} - 18{,}054 \ \text{lbf}$$
$$= 365{,}446 \ \text{lbf} \quad (370{,}000 \ \text{lbf})$$

The answer is (B).

Why Other Options Are Wrong

(A) This incorrect solution misses the MSJC code provision in Sec. 2.2.3.2 limiting in-plane tension in unreinforced masonry to zero and calculates the total load rather than the maximum applied load (total load minus self-weight).

(C) In this incorrect solution, the calculations are based on the gross section properties instead of the net section properties.

(D) This incorrect solution does not subtract the self-weight of the wall in calculating the minimum and maximum allowable applied axial loads.

SOLUTION 88

Section 13.4 of ACI 318 covers openings in slab systems. Without special analysis, openings in a two-way slab system are limited by where they occur.

Determine the width of the column strips in each direction. Column strip width on each side of a column centerline is $0.25l_1$ or $0.25l_2$, whichever is less (ACI 318 Sec. 13.2.1).

North/South Direction:

$$l_1 = 20 \ \text{ft}$$
$$l_2 = 30 \ \text{ft}$$
$$\text{column strip width} = 0.25l_1 = (0.25)(20 \ \text{ft})$$
$$= 5.0 \ \text{ft}$$
$$\text{column strip width} = 0.25l_2 = (0.25)(30 \ \text{ft})$$
$$= 7.5 \ \text{ft}$$

The lesser value controls.

East/West Direction:

$$l_1 = 30 \ \text{ft}$$
$$l_2 = 20 \ \text{ft}$$
$$\text{column strip width} = 0.25l_1 = (0.25)(30 \ \text{ft})$$
$$= 7.5 \ \text{ft}$$
$$\text{column strip width} = 0.25l_2 = (0.25)(20 \ \text{ft})$$
$$= 5.0 \ \text{ft}$$

The lesser value controls.

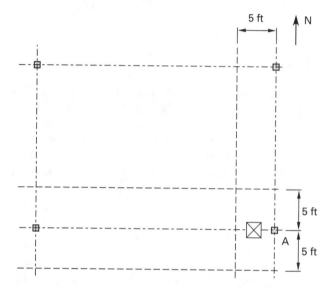

The opening falls within intersecting column strips. ACI 318 Sec. 13.4.2.2 limits openings in intersecting column strips to not more than $\tfrac{1}{8}$ the width of the column strip in either span.

$$a = b \le \left(\tfrac{1}{8}\right)(10 \ \text{ft})$$
$$= 1.25 \ \text{ft}$$

The maximum opening is

$$ab = (1.25 \ \text{ft})^2$$
$$= 1.56 \ \text{ft}^2 \quad (1.6 \ \text{ft}^2)$$

The answer is (C).

Why Other Options Are Wrong

(A) This incorrect solution assumes that since no special analysis is done, no openings are permitted.

(B) In this incorrect solution, one-half the column strip width is used instead of the full column strip width in calculating the opening size.

(D) This incorrect solution does not use the lesser value in calculating the column strip widths according to ACI 318 Sec. 13.2.1.

FAILURE ANALYSIS

SOLUTION 89

The long legs of the angle are unstiffened elements. To prevent localized buckling, AISC ASD specification Sec. B5 limits the width-to-thickness ratio so that the plate element is fully effective. For the double-angle member shown, AISC ASD Table B5.1 gives

$$\frac{b}{t} \leq \frac{76}{\sqrt{F_y}} = \frac{76}{\sqrt{36 \ \frac{\text{kips}}{\text{in}^2}}} = 12.67$$

$$\frac{b}{t} = \frac{8 \ \text{in}}{\frac{1}{2} \ \text{in}} = 16.0$$

Since $16.0 > 12.67$, the section is classified as a slender element section and must be designed in accordance with App. B of the AISC ASD.

AISC ASD App. B5.a gives the requirements for unstiffened compression elements such as the angles shown. For angles, the reduction factor, Q_s, depends on the ratio of b/t. Check if

$$\frac{b}{t} \leq \frac{155}{\sqrt{F_y}} = \frac{155}{\sqrt{36 \ \frac{\text{kips}}{\text{in}^2}}} = 25.83$$

$$\frac{b}{t} = 16.0 < \frac{155}{\sqrt{F_y}}$$

Therefore, use ASIC ASD Eq. A-B5-1.

$$Q_s = 1.340 - 0.00447 \left(\frac{b}{t}\right) \sqrt{F_y}$$
$$= 1.340 - (0.00447)(16.0)\sqrt{36}$$
$$= 0.9109$$

The allowable compressive stress also depends on

$$\frac{Kl}{r} \leq C_c'$$

$$C_c' = \sqrt{\frac{2\pi^2 E}{Q F_y}}$$

$$Q = Q_s = 0.9109$$

$$C_c' = \sqrt{\frac{2\pi^2 \left(29,000 \ \frac{\text{kips}}{\text{in}^2}\right)}{(0.9109)\left(36 \ \frac{\text{kips}}{\text{in}^2}\right)}}$$

$$= 132.1$$
$$\frac{Kl}{r} = 110 < C_c' = 132.1$$

Therefore, use AISC ASD Eq. A-B5-11.

$$F_a = \frac{Q\left(1 - \frac{\left(\frac{Kl}{r}\right)^2}{2C_c'^2}\right) F_y}{\frac{5}{3} + \frac{3\left(\frac{Kl}{r}\right)}{8C_c'} - \frac{\left(\frac{Kl}{r}\right)^3}{8C_c'^3}}$$

$$= \frac{(0.9109)\left(1 - \frac{(110)^2}{(2)(132.1)^2}\right)\left(36 \ \frac{\text{kips}}{\text{in}^2}\right)}{\frac{5}{3} + \frac{(3)(110)}{(8)(132.1)} - \frac{(110)^3}{(8)(132.1)^3}}$$

$$= 11.2 \ \text{kips/in}^2$$

Alternately, use the equation in the footnote of the Numerical Values Table 3 AISC ASD manual specification.

$$F_a = C_a Q_a Q_s F_y$$

$$\frac{\left(\frac{Kl}{r}\right)}{C_c'} = \frac{110}{132.1} = 0.833$$

By interpolation, find C_a in AISC ASD Table 3 to be 0.343.

$$F_a = C_a Q_a Q_s F_y$$
$$= (0.343)(1.0)(0.9109)\left(36 \ \frac{\text{kips}}{\text{in}^2}\right)$$
$$= 11.2 \ \text{kips/in}^2$$

The answer is (C).

Why Other Options Are Wrong

(A) This incorrect solution uses the sum of the widths of the two angles in calculating the width/thickness ratio. The correct solution considers the slenderness of each angle individually.

(B) This incorrect solution solves the problem correctly but identifies the wrong units for allowable stress.

(D) This incorrect solution does not take slenderness into account when determining the allowable stress.

SOLUTION 90

Frequent variations or reversals of stress can cause fatigue. Appendix Chapter K of the AISC ASD manual gives requirements for fatigue loading.

If the number of loading cycles exceeds 20,000 over the lifetime of the member, fatigue must be considered. Using the information in the footnotes to AISC ASD Table A-K4.1, determine that 200 applications every day is approximately equivalent to 2×10^6 loading cycles over the lifetime of this structure. Loading Condition 4 applies.

From AISC ASD Fig. A-K4.1, determine that a bolted end connection is shown in AISC ASD Examples 8 and 9. From AISC ASD Table A-K4.2, under General Condition of Mechanical Fasteners, determine that for a bolted connection, the Stress Category is B.

Use AISC ASD Table A-K4.3 to determine the allowable stress range to be 16 kips/in^2 for Stress Category B and Loading Condition 4.

Determine the range of loads.

$$D + L = 20 \text{ kips} + 50 \text{ kips}$$
$$= 70 \text{ kips (tension)}$$
$$D + L = 20 \text{ kips} - 10 \text{ kips}$$
$$= 10 \text{ kips (tension)}$$

The allowable tensile stress is

$$F_t = 0.6F_y = (0.6)\left(36 \, \frac{\text{kips}}{\text{in}^2}\right)$$
$$= 21.6 \text{ kips/in}^2$$

Estimate the channel size.

$$A_{\text{req}} = \frac{P_{\text{max}}}{F_t} = \frac{70 \text{ kips}}{21.6 \, \frac{\text{kips}}{\text{in}^2}}$$
$$= 3.24 \text{ in}^2$$

Try a C8 × 11.5.

$$A_{\text{prov}} = 3.38 \text{ in}^2$$
$$f_{t,\text{max}} = \frac{P_{\text{max}}}{A} = \frac{70 \text{ kips}}{3.38 \text{ in}^2}$$
$$= 20.7 \text{ kips/in}^2$$
$$f_{t,\text{min}} = \frac{P_{\text{min}}}{A} = \frac{10 \text{ kips}}{3.38 \text{ in}^2}$$
$$= 2.96 \text{ kips/in}^2$$

The actual stress range is

$$f_{\text{sr}} = 20.7 \, \frac{\text{kips}}{\text{in}^2} - 2.96 \, \frac{\text{kips}}{\text{in}^2}$$
$$= 17.7 \text{ kips/in}^2$$

The allowable stress range is

$$F_{\text{sr}} = 16 \text{ kips/in}^2 < 17.7 \text{ kips/in}^2 \quad \text{[no good]}$$

Try a channel with a larger cross section, such as a C7 × 12.25.

$$A_{\text{prov}} = 3.60 \text{ in}^2$$
$$f_{t,\text{max}} = \frac{P_{\text{max}}}{A} = \frac{70 \text{ kips}}{3.60 \text{ in}^2}$$
$$= 19.4 \text{ kips/in}^2$$
$$f_{t,\text{min}} = \frac{P_{\text{min}}}{A} = \frac{10 \text{ kips}}{3.60 \text{ in}^2}$$
$$= 2.78 \text{ kips/in}^2$$

The actual stress range is

$$f_{\text{sr}} = 19.4 \, \frac{\text{kips}}{\text{in}^2} - 2.78 \, \frac{\text{kips}}{\text{in}^2}$$
$$= 16.6 \text{ kips/in}^2$$

The allowable stress range is

$$F_{\text{sr}} = 16 \text{ kips/in}^2 < 16.6 \text{ kips/in}^2 \quad \text{[no good]}$$

Try a C6 × 13.

$$A_{\text{prov}} = 3.83 \text{ in}^2$$
$$f_{t,\text{max}} = \frac{P_{\text{max}}}{A} = \frac{70 \text{ kips}}{3.83 \text{ in}^2}$$
$$= 18.3 \text{ kips/in}^2$$
$$f_{t,\text{min}} = \frac{P_{\text{min}}}{A} = \frac{10 \text{ kips}}{3.83 \text{ in}^2}$$
$$= 2.61 \text{ kips/in}^2$$

The actual stress range is

$$f_{\text{sr}} = 18.3 \, \frac{\text{kips}}{\text{in}^2} - 2.61 \, \frac{\text{kips}}{\text{in}^2}$$
$$= 15.7 \text{ kips/in}^2$$

The allowable stress range is

$$F_{\text{sr}} = 16 \text{ kips/in}^2 > 15.7 \text{ kips/in}^2 \quad \text{[OK]}$$

Use a C6 × 13.

The answer is (B).

Why Other Options Are Wrong

(A) This incorrect solution adds the alternating live loads together instead of treating them as separate loading conditions.

(C) This incorrect solution ignores the effects of fatigue.

(D) This incorrect solution misreads Table A-K4.3 for Allowable Stress Range for Category B′ (12 kips/in²) instead of for Category B.

DESIGN CRITERIA

SOLUTION 91

ASTM specification C652 is the *Standard Specification for Hollow Brick (Hollow Masonry Units Made from Clay or Shale)*. Therefore, the equation for the modulus of elasticity for clay masonry must be used.

The modulus of elasticity of clay masonry is given in Sec. 1.8.2.2.1 of the MSJC code as

$$E_m = 700 f'_m$$
$$= (700)\left(4500 \ \frac{\text{lbf}}{\text{in}^2}\right)$$
$$= 3,150,000 \ \text{lbf/in}^2 \quad (3.2 \times 10^6 \ \text{lbf/in}^2)$$

The answer is (C).

Why Other Options Are Wrong

(A) This incorrect solution misreads the specified compressive strength of masonry given in the problem statement as the compressive strength of the masonry unit. It uses Part 1, Table 1 of the MSJC specification to determine the compressive strength of 4500 lbf/in² clay masonry laid in Type S mortar to be 1859 lbf/in² and uses this value in calculating the modulus of elasticity.

(B) This incorrect solution mistakenly identifies that the unit is made from concrete and misreads the specified compressive strength of masonry given in the problem statement as the compressive strength of the masonry unit. It uses MSJC specification Part 1, Table 2 and the equation for modulus of elasticity, E_m, for concrete masonry units.

(D) This incorrect solution uses the equation for the modulus of elasticity of concrete masonry instead of that for clay masonry.

SOLUTION 92

Section 1609 of the IBC covers wind design. This section requires that wind loads be determined using Sec. 6 of ASCE 7 unless the simplified wind load method found in IBC Sec. 1609.6 is used. Section 1609.6 limits the use of the simplified method to enclosed buildings. Therefore, a tank cannot be designed using this approach.

ASCE 7 Sec. 6 offers three methods of analysis. Method 1 is a simplified method that applies only to buildings. Method 3 is a wind tunnel procedure. For the tank in this case, use the analytical procedure of Method 2 in Sec. 6.5 of ASCE 7.

The design wind force for the tank is given in ASCE 7 Sec. 6.5.13 as

$$F = q_z G C_f A_f \quad \text{[ASCE 7 Eq. 6-25]}$$

The velocity pressure, q_z, is evaluated at the height of the centroid of the projected area, A_f, of the shape. For a 50 ft high tank with a 20 ft diameter, the height of the centroid is

$$z = h - \frac{D}{2} = 50 \ \text{ft} - \frac{20 \ \text{ft}}{2} = 40 \ \text{ft}$$

The velocity pressure is given as

$$q_z = 0.00256 K_z K_{zt} K_d \text{v}^2 I_w \quad \text{[ASCE 7 Eq. 6-15]}$$

The velocity pressure coefficient is a function of exposure category. From IBC Sec. 1609.4, determine that a West Coast shoreline exposure is Exposure D. Using ASCE 7 Table 6-3, determine the velocity pressure coefficient at 40 ft to be

$$K_z = 1.22$$

The topographic factor, K_{zt}, given in ASCE 7 Sec. 6.5.7.2 applies to locations that have abrupt changes in general topography. Since the landscape is flat in this case, K_{zt} does not apply.

The wind directionality factor for round tanks is given in ASCE 7 Table 6-4 as

$$K_d = 0.95$$

The basic wind speed, v, is given as 85 mph. (See also IBC Fig. 1609.)

The importance factor, I_w, is given in IBC Table 1604.5. For water tanks in building Categories III and IV,

$$I_w = 1.15$$

The velocity pressure is

$$q_z = (0.00256)(1.22)(1.0)(0.95)\left(85 \ \frac{\text{mi}}{\text{hr}}\right)^2 (1.15)$$
$$= 24.7 \ \text{lbf/ft}^2$$

The gust effect factor, G, for rigid structures is given in ASCE 7 Sec. 6.5.8.1 as 0.85.

The net force coefficients for tanks are found in ASCE 7 Fig. 6-19 and are based on the ratio of the structure's height to diameter and on the velocity pressure. In this case,

$$h = 50 \ \text{ft}$$
$$D = 20 \ \text{ft}$$
$$\frac{h}{D} = \frac{50 \ \text{ft}}{20 \ \text{ft}} = 2.5$$
$$D\sqrt{q_z} = (20 \ \text{ft})\sqrt{24.7 \ \frac{\text{lbf}}{\text{ft}^2}} = 99.4$$

For a moderately smooth tank, interpolate ASCE 7 Fig. 6-19 to determine

$$C_f = 0.54$$

The projected area of the circular tank is

$$A_f = \pi \left(\frac{D^2}{4} \right) = \pi \left(\frac{(20 \text{ ft})^2}{4} \right) = 314.2 \text{ ft}^2$$

The design wind force for the tank is

$$\begin{aligned} F &= q_z G C_f A_f \\ &= \left(24.7 \; \frac{\text{lbf}}{\text{ft}^2} \right) (0.85)(0.54)(314.2 \text{ ft}^2) \\ &= 3560 \text{ lbf} \end{aligned}$$

The answer is (C).

Why Other Options Are Wrong

(A) This incorrect solution determines the velocity pressure (24.7 lbf/ft^2), not the design wind force.

(B) This solution mistakenly identifies the exposure category as C, which applies to terrain that is flat and generally open. Although the terrain at this location is likely flat and open, the correct exposure is D because the tank is located near a large body of water. The velocity pressure coefficient, K_z, is incorrectly determined to be 1.04.

(D) This incorrect solution reads ASCE 7 Table 6-3 for a 50 ft height instead of the height to the centroid of 40 ft and determines K_z to be 1.27.

SOLUTION 93

Sections 1617.3 and 1617.5.4 of the IBC cover story drift limitations. Only buildings in Seismic Design Category A do not need to comply with these requirements (see IBC Sec. 1616.4). Therefore, statement C is false.

The answer is (C).

Why Other Options Are Wrong

(A) According to IBC Ch. 16, the design lateral seismic force is directly proportional to the building weight. For example, IBC Sec. 1617.5 gives the following equation for seismic base shear.

$$V = \left(\frac{1.2 S_{DS}}{R} \right) W \quad \text{[IBC Eq. 16-56]}$$

In this equation, W is the effective seismic weight, which includes the total dead load. As W increases, so does the total design base shear, V. Statement A is true.

(B) Seismic Design Category is based on Seismic Use Group and the design spectral response coefficients, S_{DS} and S_{DI}.

According to IBC Sec. 1616.2, an apartment building is classified as Seismic Use Group II.

Use IBC Table 1616.3(1) to determine that S_{DS} is between $0.33g$ and $0.50g$, knowing the Seismic Design Category (C) and Use Group (II).

(D) IBC Sec. 1615.1.1 covers site class definitions. The exception in this section states, "When the soil properties are not known in sufficient detail to determine the site class, Site Class D shall be used..." Statement D is true.

SOLUTION 94

Section 1805 of the IBC, Footings and Foundations, includes prescriptive requirements for foundation walls.

Section 1805.5.1.1 of the IBC specifies that the minimum thickness of foundation walls must not be less than the thickness of the wall supported. Therefore, the minimum thickness of the foundation wall must be equal to the concrete masonry wall thickness, or 12 in.

Verify that 12 in is adequate based on the soil lateral load. Begin by determining the soil class from IBC Table 1610.1, which states, "Poorly graded clean sands; sand-gravel mixes" have a unified soil classification of SP and an active design lateral soil load of 30 lbf/ft^2-ft.

IBC Table 1805.5(1) specifies the minimum thickness for plain concrete and plain masonry foundation walls. The minimum thickness depends upon the wall height and the height of the unbalanced backfill. The wall height is 8 ft. The height of unbalanced backfill is 8 ft minus 4 ft, which equals 4 ft. Knowing these values and the unified soil classification, SP, IBC Table 1805.5(1) requires a minimum wall thickness of 7.5 in.

The answer is (D).

Why Other Options Are Wrong

(A) This incorrect solution overlooks the requirements of IBC Sec. 1805.5.1.1 for minimum thickness equal to the supported wall thickness.

(B) This incorrect solution overlooks the requirements of IBC Sec. 1805.5.1.1 for minimum thickness equal to the supported wall thickness and uses the portion of IBC Table ~~1610.1~~ for plain masonry walls instead of plain concrete walls. *1805.5(1)*

(C) This incorrect solution miscalculates the height of unbalanced backfill in IBC Table ~~1610.1~~ as 8 ft instead of 4 ft and overlooks the requirements of IBC Sec. *1805.5(1)*

1805.5.1.1 for minimum thickness equal to the supported wall thickness.

SOLUTION 95

Section 1607.11 of the IBC states that roofs must be designed for the appropriate live loads. This section also contains requirements for minimum roof live loads. Minimum roof live loads are between 12 lbf/ft^2 and 20 lbf/ft^2 (IBC Eq. 16-24). In this case, the given design roof live load (40 lbf/ft^2) exceeds the minimum roof load.

The tributary area of the column is

$$A_t = (20 \text{ ft})(15 \text{ ft}) = 300 \text{ ft}^2$$

The live load on the column is

$$P = LA_t = \left(40 \; \frac{\text{lbf}}{\text{ft}^2}\right)(300 \text{ ft}^2)\left(\frac{1 \text{ kip}}{1000 \text{ lbf}}\right)$$
$$= 12.0 \text{ kips} \quad (12 \text{ kips})$$

The answer is (C).

Why Other Options Are Wrong

(A) This incorrect solution does not recognize the roof live load calculated by IBC Sec. 1607.11.2 as a minimum load and uses IBC Sec. 1607.11.2.1 to calculate a reduced live load based on the 20 lbf/ft^2 given in this section.

(B) This incorrect solution calculates the reduction factors given in IBC Sec. 1607.11.2.1 and applies them to the design live load of 40 lbf/ft^2.

(D) This incorrect solution calculates the total load, not the live load.

SOLUTION 96

Section 1607 of the IBC covers live loads. Roof live loads are ~~permitted to be reduced~~ *determined* in accordance with IBC Sec. 1607.11.2 depending on area supported and slope of the roof. The reduced design live load is calculated as

$$L_r = \left(20 \; \frac{\text{lbf}}{\text{ft}^2}\right) R_1 R_2 \quad [\text{IBC Eq. 16-24}]$$

The area supported by an interior girder is

$$A_t = bl = (50 \text{ ft})(30 \text{ ft}) = 1500 \text{ ft}^2$$

The reduction factor, R_1, for tributary areas 600 ft^2 or greater is 0.6 (IBC Eq. 16-27). A flat roof has a slope factor, F, of 0. Therefore, the reduction factor, R_2, is 1 (IBC Eq. 16-28).

The reduced live load on the girder is

$$L_r = \left(20 \; \frac{\text{lbf}}{\text{ft}^2}\right)(0.6)(1) = 12.0 \text{ lbf/ft}^2$$

The live load on the girder is

$$P_L = L_r A = \left(12.0 \; \frac{\text{lbf}}{\text{ft}^2}\right)\left(\frac{1 \text{ kip}}{1000 \text{ lbf}}\right)(1500 \text{ ft}^2)$$
$$= 18.0 \text{ kips} \quad (18 \text{ kips})$$

The answer is (B).

Why Other Options Are Wrong

(A) This incorrect solution applies the alternate floor live load reduction found in IBC Sec. 1607.9.2 to the roof live load and calculates the reduction incorrectly. The reduced live load should be calculated as $(1 - R)$ times the design live load, not R times the load, as is done in this solution.

(C) This incorrect solution applies the alternate floor live load reduction found in IBC Sec. 1607.9.2 to the roof live load and makes a mathematical error in calculating the area supported.

The area supported by an interior girder is

$$A = bl = (50 \text{ ft})(30 \text{ ft})$$
$$= 150 \text{ ft}^2 \quad [\text{this is not correct}]$$

(D) This incorrect solution does not ~~reduce the roof live load~~ *use IBC Eq. 16-24 to calculate the roof live load*

$$P_L = LA = \left(20 \; \frac{\text{lbf}}{\text{ft}^2}\right)\left(\frac{1 \text{ kip}}{1000 \text{ lbf}}\right)(1500 \text{ ft}^2) = 30 \text{ kips}$$

SOLUTION 97

A roof member that does not support a ceiling is limited by Table 1604.3 of the IBC to a total load deflection of $L/120$.

$$\Delta = \frac{L}{120} = \left(\frac{60 \text{ ft}}{120}\right)\left(12 \; \frac{\text{in}}{\text{ft}}\right)$$
$$= 6.0 \text{ in} \quad (6 \text{ in})$$

The answer is (D).

Why Other Options Are Wrong

(A) This incorrect solution does not convert the length of the member to inches in calculating the deflection limit. The units do not work out.

(B) This incorrect solution uses the deflection limit for live load with nonplastered ceilings, $L/240$, instead of the deflection limit for total load with no ceiling.

(C) This incorrect solution uses the total load deflection limit for nonplastered ceilings, $L/180$, instead of the deflection limit for no ceiling.

SOLUTION 98

Section 1.14 of the MSJC code contains the quality assurance requirements for masonry structures. There are three levels of quality assurance, depending on the type of facility (essential or nonessential) and the method of design.

A residence is a nonessential facility. Since this building was designed using the strength design provisions, it was *not* designed using MSJC Chs. 5, 6, or 7, which cover empirical design, veneer, and glass unit masonry. MSJC Sec. 1.14.3 requires such buildings to comply with MSJC Table 1.14.2 of the MSJC code.

Table 1.14.2 of the MSJC code (Level 2 quality assurance) contains, among others, the following inspection requirements.

Prior to grouting, verify

- grout space
- grade and size of reinforcement
- placement of reinforcement
- proportions of site-prepared grout
- construction of mortar joints

Verify that the placement of grout is in compliance.

Observe preparation of grout specimens, mortar specimens, and/or prisms.

Verify compressive strength of masonry, f'_m, prior to construction.

Statements I and III are true for Level 2 quality assurance. Statement II is a provision of Level 3 quality assurance that requires *continuous* verification of grout placement. Statement IV is a provision of Level 3 quality assurance that requires continual verification of f'_m throughout construction (every 5000 ft^2).

The answer is (B).

Why Other Options Are Wrong

(A) This incorrect solution correctly identifies statement I as true, but statement II is false. Statement I is true because, for this building, Level 2 quality assurance is required. Level 2 quality assurance includes, among others, the requirement that prior to grouting, placement of reinforcement must be verified.

Continuous verification of placement of reinforcement (statement II) is a provision of Level 3 quality assurance.

(C) This incorrect solution identifies statements I and III as true, but does not recognize statement II as not applicable to Level 2 quality assurance.

(D) This incorrect solution mistakenly applies the provisions for buildings designed using Chs. 5, 6, or 7 of the MSJC code. Strength design is found in Ch. 3 of the MSJC code.

SOLUTION 99

The tabulated allowable bending design values must be multiplied by the beam stability factor, C_L, as specified in NDS Sec. 3.3.3. Statement I is false.

The size factor does not apply to structural glulam timber. The adjustments to design values for structural glulam timber are given in NDS Sec. 5.3. Statement II is true.

NDS Sec. 6.3.5 specifies that the tabulated allowable design values for round timber piles include adjustments to compensate for the strength reductions due to steam conditioning or boultonizing. The tabulated design values for untreated timber piles must be multiplied by the untreated factor, C_u. Statement III is false.

The group action factor does not apply to all types of wood connections. Timber rivets, metal plate connections, and spikes are among those connections to which the group action factor does not apply (NDS Sec. 10.3.1). Statement IV is false.

The answer is (D).

(A) This incorrect solution only identifies the first false statement. Statements III and IV are also false.

(B) This incorrect solution identifies the only true statement rather than those that are not true.

(C) This incorrect solution correctly identifies statements I and IV as false but misreads NDS Sec. 6.3.5 to indicate that the untreated factor is already incorporated into the design values.

SOLUTION 100

The load combinations to be considered are given in IBC Sec. 1605.3.

$$D$$
$$D + L_r$$
$$D + W + L_r$$
$$0.6D + W$$

IBC Sec. 1605.3.1.1 also contains the provision that the combined effect of two or more variable loads can be multiplied by 0.75 and added to the effect of the dead load.

$$D + 0.75(W + L_r)$$

Calculate the dead load on the roof deck.

$$\text{roof deck} = 3 \text{ lbf/ft}^2$$
$$\text{rigid insulation} = 3 \text{ lbf/ft}^2$$
$$\text{felt and gravel} = 5 \text{ lbf/ft}^2$$
$$\text{TOTAL} = 11 \text{ lbf/ft}^2$$

The live load for the roof deck is given as 20 lbf/ft^2.

Determine the wind load using IBC Sec. 1609. The building meets the requirements of the simplified provisions for low-rise buildings and can be designed in accordance with IBC Sec. 1609.6.

The basic wind speed, v, is given as 100 mph.

From IBC Table 1604.5, determine that a ~~fire station~~ *warehouse* is Building Category ~~IV~~ *II* and the importance factor, I_w, is ~~1.15.~~ *1.00*

From IBC Sec. 1609.4, determine the exposure category to be B. From IBC Table 1609.6.2.1(4), determine the height and exposure adjustment coefficient, λ, for a 30 ft roof height and Exposure B to be 1.00.

The simplified design wind pressure for the MWFRS is

$$p_s = \lambda I_w p_{s30} \quad \text{[IBC Eq. 16-34]}$$

From IBC Fig. 1609.6.2.1, determine the roof to be classified for Zones E, F, G, and H. These zones account for localized higher pressures.

Using IBC Table 1609.6.2.1(1), determine the maximum and minimum simplified pressures, p_{s30}, for a 100 mph wind speed and a flat roof for Zones E, F, G, and H. The maximum and minimum pressures are for Zones E and H, respectively.

$$p_{s30,\text{max}} = -19.1 \text{ lbf/ft}^2$$
$$p_{s30,\text{min}} = -8.4 \text{ lbf/ft}^2$$

Calculate the maximum and minimum wind pressures.

$$p_s = \lambda I_w p_{s30}$$
$$p_{s,\text{max}} = (1.00)(\underset{1.00}{\cancel{1.15}})\left(-19.1 \frac{\text{lbf}}{\text{ft}^2}\right) = \underset{-19.1}{\cancel{-22.0}} \text{ lbf/ft}^2$$
$$p_{s,\text{min}} = (1.00)(\underset{1.00}{\cancel{1.15}})\left(-8.4 \frac{\text{lbf}}{\text{ft}^2}\right) = \underset{-8.4}{\cancel{-9.7}} \text{ lbf/ft}^2$$

Determine the worst-case load.

By inspection determine that dead load alone is not the critical combination.

$$D + L_r = 11 \frac{\text{lbf}}{\text{ft}^2} + 20 \frac{\text{lbf}}{\text{ft}^2} = 31 \text{ lbf/ft}^2$$

$D + W + L_r$:

$$D + p_{s,\text{max}} + L_r = 11 \frac{\text{lbf}}{\text{ft}^2} \overset{19.1}{\underset{\underset{=9.0\text{ lbf/ft}^2}{11.9}}{\cancel{-22.0}}} \frac{\text{lbf}}{\text{ft}^2} + 20 \frac{\text{lbf}}{\text{ft}^2}$$

$$D + p_{s,\text{min}} + L_r = 11 \frac{\text{lbf}}{\text{ft}^2} \overset{8.4}{\cancel{-9.7}} \frac{\text{lbf}}{\text{ft}^2} + 20 \frac{\text{lbf}}{\text{ft}^2}$$
$$\underset{22.6}{= \cancel{21} \text{ lbf/ft}^2}$$

$0.6D + W$:

$$0.6D + p_{s,\text{max}} = (0.6)\left(11 \frac{\text{lbf}}{\text{ft}^2}\right) \overset{19.1}{\cancel{-22.0}} \frac{\text{lbf}}{\text{ft}^2}$$
$$\underset{-12.5}{} \underset{-15 \text{ lbf/ft}^2}{=}$$

$$0.6D + p_{s,\text{min}} = (0.6)\left(11 \frac{\text{lbf}}{\text{ft}^2}\right) \overset{8.4}{\cancel{-9.7}} \frac{\text{lbf}}{\text{ft}^2}$$
$$\underset{-1.8}{= \cancel{-3.1} \text{ lbf/ft}^2}$$

$D + 0.75(W + L_r)$:

$$D + 0.75(p_{s,\text{max}} + L_r) = 11 \frac{\text{lbf}}{\text{ft}^2} + (0.75)$$
$$\times \left(\overset{19.1}{\cancel{-22.0}} \frac{\text{lbf}}{\text{ft}^2} + 20 \frac{\text{lbf}}{\text{ft}^2}\right)$$
$$\underset{11.7}{= \cancel{9.5} \text{ lbf/ft}^2}$$

$$D + 0.75(p_{s,\text{min}} + L_r) = 11 \frac{\text{lbf}}{\text{ft}^2} + (0.75)$$
$$\times \left(\overset{8.4}{\cancel{-9.7}} \frac{\text{lbf}}{\text{ft}^2} + 20 \frac{\text{lbf}}{\text{ft}^2}\right)$$
$$\underset{19.7}{= \cancel{19} \text{ lbf/ft}^2}$$

The critical load case is that which is maximum, $D + L_r$. The critical design load is 31 lbf/ft^2.

The answer is (C).

Why Other Options Are Wrong

(A) This incorrect solution gives the largest uplift load from the $0.6D + W$ load combination (-15 lbf/ft^2).

(B) This incorrect solution gives the maximum uplift pressure (-22 lbf/ft^2) and does not consider the load combinations.

(D) This incorrect solution uses the load combinations for strength design instead of those for allowable stress.

The critical design case is $1.2D + 1.6L_r$.

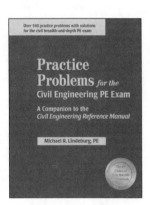
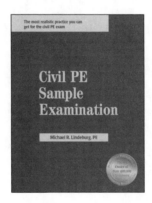